AF352334

STEREOLOGY
AND MORPHOMETRY
IN ELECTRON MICROSCOPY

STEREOLOGY AND MORPHOMETRY IN ELECTRON MICROSCOPY

problems and solutions

Edited by

albrecht reith

Laboratory for Electron Microscopy and Morphometry, The Norwegian Radium Hospital, Oslo, Norway

terry m. mayhew

Department of Anatomy, Marischal College, Aberdeen, Scotland

With a Foreword by

ewald r. weibel

⬤hemisphere publishing corporation

New York   Washington   Philadelphia   London

STEREOLOGY AND MORPHOMETRY IN ELECTRON MICROSCOPY: Problems and Solutions

1 2 3 4 5 6 7 8 9 0   B R B R   8 9 8

This book was set in Times Roman by The Sheridan Press. The editors were Sandra Tamburrino and Mary Prescott.
Braun-Brumfield, Inc. was printer and binder.

Library of Congress Cataloging in Publication Data

Stereology and morphometry in electron microscopy: problems and
solutions / edited by Albrecht Reith, Terry M. Mayhew.
    p.   cm.—(An Ultrastructural pathology publication series.
ISSN 0730-6482)
  Bibliography: p.
  Includes index.
  1. Electron microscopy—Technique. 2. Stereology. I. Reith,
Albrecht. II. Mayhew, Terry M. III. Series.
RB43.5.S74 1988
616.07'58—dc19   87-35365 CIP
ISBN 0-89116-623-8
ISSN 0730-6482

To the students and teachers of the annual courses in morphometry and stereology, whose enthusiasm and many fruitful discussions have made this book possible.

contents

CONTENTS

contributors

Steinar Aase, MD, PhD
Institute of Pathology
Rikshospitalet
0027 Oslo 1
Norway

Luis M. Cruz-Orive, PhD
Department of Anatomy
University of Bern
P.O. Box 139
CH-3000 Bern 9
Switzerland

Didima M. G. de Groot, PhD
Medical Biological Laboratory TNO
P.O. Box 45
2280 AA Rijswijk
The Netherlands

O. Götzsche, MD
Stereological Research Laboratory
University Institute of Pathology and
    2nd University Clinic of Internal
    Medicine
Institute of Experimental Clinical
    Research
University of Århus
DK-8000 Århus
Denmark

H. J. G. Gundersen, MD
Stereological Research Laboratory
University Institute of Pathology and
    2nd University Clinic of Internal
    Medicine
Institute of Experimental Clinical
    Research
University of Århus
DK-8000 Århus
Denmark

Jostein Halgunset, MD
Department of Pathology
University of Trondheim
N-7000 Trondheim
Norway

E. B. Jensen, DSc
Department of Theoretical Statistics
Institute of Mathematics
University of Århus
DK-8000 Århus
Denmark

Michel Kraemer, PhD
Laboratoire de Biologie du
    Développement
UER Biomedicale
Université Paris XIII
93012 Bobigny-Cedex
France

Terry M. Mayhew, PhD
Department of Anatomy
Marischal College
University of Aberdeen
Aberdeen AB9 1AS
Scotland

Carolyn Middleton, BSc
Department of Anatomy
Marischal College
University of Aberdeen
Aberdeen AB9 1AS
Scotland

Andreas O. Myking, MD
The Gade Institute
Department of Pathology
5016 Haukeland sykehus
Norway

R. Østerby, MD
Diabetes Research Laboratory
University Institute of Pathology
Århus Kommunehospital
DK-8000 Århus
Denmark

Kristian Ree, MD
Department of Dermatology
Ullevål Hospital
0407 Oslo 4
Norway

Jon B. Reitan, MD
National Institute of Radiation
    Hygiene
P.O. Box 55
N-1345 Østerås
Norway

Albrecht Reith, MD, PhD
Laboratory for Electron Microscopy
    and Morphometry
Institute for Cancer Research
The Norwegian Radium Hospital
Montebello
0310 Oslo 3
Norway

Jean Paul Rigaut, MD, ScD, PhD
INSERM U.263
Université Paris 7
Tour 53
2 Place Jussieu
75251 Paris Cedex 05
France

G. A. Ross, BSc
Department of Anatomy
Marischal College
University of Aberdeen
Aberdeen AB9 1AS
Scotland

Jeannie Vassy, BSc
Laboratoire de Biologie du
    Développement
UER Biomedicale
Université Paris XIII
93012 Bobigny-Cedex
France

foreword

It is now over a quarter of a century since the terms *stereology* and *morphometry* surfaced, and still they are something like "scarecrows" for many microscopists—the most avant-garde electron microscopists included. Why? Because stereology and morphometry deal with numbers and not so much with pictures; or perhaps one should say with numbers derived from pictures. Now the science of handling numbers, of making sense of them, is mathematics. And biologists, particularly morphologists, commonly experience a deep sense of horror or shock when they are asked to think mathematically, that is to say, to make an effort to reduce part of the wealth of information contained in their pictures to some simple quantitative bits of information. I daresay it is not beauty that is lost in this process. The picture goes undamaged and can safely be hung on the wall. It is rather beauty that is gained, but beauty of a different kind: one that allows us to discover some of the laws of nature that we humans—with our limited capabilities—need to single out and study one by one before we can put them back together. But in the end, understanding nature, the conditions of life in particular, is an attainable goal only if we are willing to sort out the many facets of nature, and of life, and to put them into higher perspective. Morphometry is but one approach, and stereology but one of the tools that help us to strive for this goal.

In recent years stereology has evolved greatly. On the one hand, the theory behind it has become rather abstract, difficult to grasp for the nonspecialist. On the other hand, however, the actual methods of stereology have become simpler and easier to apply—provided one abides by the rules set by the theoreticians and is prepared to trust them as one trusts the engineer who designed the electron microscope one uses. Originally, stereology was defined as "a set of mathematical methods relating three-dimensional parameters of structure to two-dimensional measurements obtained on sections"; today it appears more appropriate to define it simply, and more practically, as the theory of sampling spatial structures for measurement.

This new trend is felt throughout this book. Its articles are derived from the Scandinavian course on stereology, which has an established record for teaching a difficult topic so as to reach a varied audience. It is not a "cookbook" but a useful guide for those who are willing to undertake the job of understanding what is behind the nice pictures their electron microscopes produce: the laws that govern the design of living structures—order and proportions that make sense.

Ewald R. Weibel
Professor of Anatomy
University of Bern

preface

It is encouraging to see how the use of morphometric techniques in all morphology-linked fields of science has developed in recent years, especially in pathology and electron microscopy. This is particularly encouraging for those of us who have over the past 10 years arranged and lectured at the annual Scandinavian courses in morphometry and stereology. From the start in 1977, these courses have been supported by the Norwegian Society of Pathology and the Scandinavian Society for Electron Microscopy, and more than 200 people have been trained in the techniques. In fact, it can be said that this book is based on the experiences gained during these annual courses, and the contributors have all been either members of the faculty or students, both equally enthusiastic.

The increasing interest in the techniques shown by pathologists is to some extent related to an aspect of the nature of disease as described by Forbus (1943) in his book *Reaction to Injury*: "Disease is not the imposition of new, different structures and functions, but simply the *quantitative* alteration—increase or decrease—of existing pathways." But the main interest stems from the realization that subjective morphological descriptions, such as "a moderate increase in mitochondria volume was observed," are inadequate and that morphometric techniques offer objectivity and reproducibility, as in the statement "the volume fraction (or density) of mitochondria was 20% in hepatocyte cytoplasm."

Despite their finely developed ability to interpret morphological patterns by eye and brain, there are some things pathologists cannot assess by this means, mainly *quantitative* differences. Volume or number increases must be on the order of 30–50% to be recognizable, and the increases in membrane, that is, surface density, have to be even higher. However, in toxicological studies, for example, a constant 10–20% increase of the endoplasmic reticulum may be an important observation, but this will be lost in any routine electron microscopy investigation. By applying quantitative methods to electron microscopy, numerical data will be obtained (an advantage in itself) that will bring electron microscopy to the level of biochemistry and physiology, in which quantitation was achieved decades ago.

That quantitation (by morphometry and stereology) in pathology and electron microscopy has not been applied earlier and to a larger extent is partly a consequence of the very theoretical way in which the main textbooks and reviews have presented the technique. In addition, data collecting and computing have been considered to be too time consuming. By the use of cheap desk computers and pocket calculators, this time obstacle has been overcome since one can count directly into the machines, which, once programmed, perform the required calculations (e.g., estimating mean values and standard errors and performing the statistical tests automatically).

Based on our experience, both in our own work and as advisers to many colleagues and course participants, the most helpful advice we can give to newcomers is to start with easy, uncomplicated parameters such as volume and surface estimations. The actual measurements and countings on the section or micrograph by the point and/or line grid method are not difficult. But special consideration must be given to sampling at all levels, that is, animals, blocks, sections, and micrographs. The difficulties and likelihood of error and how to overcome them will be dealt with in the introductory chapter, where we will also describe parameters other than volume and surface densities.

The later chapters deal with specific problems of common interest to all who practice morphometric and stereologic techniques. An introduction in every chapter describes the methodological aspect for which the example was chosen. Learning by example is, we hope, an easier way to become familiar with morphometric and stereologic methods than by going back to textbooks. This, at least, was the intention of the editors.

Albrecht Reith
Terry M. Mayhew

acknowledgments

We are indebted to the Nordic Council of Ministers, which through its Secretariat for Nordic Cultural Exchange has supported the annual courses in morphometry and stereology. We thank the Norwegian Radium Hospital and the British Council for contributing toward travel expenses in connection with the preparation of this book. Helpful discussions with Professor Herbert Helander of Gothenburg, Hans Jørgen G. Gundersen of Århus, and Luis M. Cruz-Orive of Bern are particularly acknowledged.

STEREOLOGY
AND MORPHOMETRY
IN ELECTRON MICROSCOPY

chapter 1

introducing basic principles and methods of stereology and morphometry

T. M. Mayhew and A. Reith

INTRODUCTION

Progress in electron microscopy has been marked by advances in new instrumentation or techniques of specimen preparation (e.g., use of X-ray detectors, perfusion fixation, immunocytochemistry). Quite a different and new technique for analyzing ultrastructural changes is the quantitative approach of assessing morphological structures in terms of number, surface, or volume densities. During the past two decades, research in light microscopy and electron microscopy has involved increasing use of morphometric, particularly stereological, methods.

After the pioneering years of qualitative observation, electron microscopists came to appreciate the benefits of quantification. They observed that cells are compartmentalized and that subcellular compartments (mitochondria, endoplasmic reticulum, lysosomes, etc.) change with age, disease, and experimental manipulation. With the development of quantitative methods for biochemical and physiological studies, correlation between structure and function came to demand the use of quantitative morphological methods.

Some Definitions

Morphometry, as the name implies, is simply the measurement of form. For the anthropometrist, measurements may be made on the *intact* individual, but the electron microscopist must first generate *representations* of cells by sectioning fixed and embedded material. In so doing, structural integrity is preserved in only two dimensions because the third dimension is sacrificed to optimal lateral resolution. For the electron microscopist, then, measurements are made on sections. If the section thickness is negligible in comparison to the size of the object under investigation, then the measurements will yield essentially *planar* morphometric data.

Stereology may be regarded as one aspect of morphometry involving the use of mathematical relationships to define three-dimensional quantities from measurements made on (ideally) two-dimensional representations (e.g., thin sections). The relationships are based on the reasonings of geometric probability and statistics. Fortunately, these relationships are numerically simple and easy to apply, so the potential user has *no excuse* for being discouraged from applying them.

Some References

This chapter cannot do justice to the morphometric/stereological methods available. This introduction must therefore be supplemented by additional reading. Therefore, the reader is referred to the principal articles and books.[1-16]

ACQUIRING DATA FROM THIN SECTIONS

Measurements of structural features appearing on sections (or micrographs) can be obtained by superimposing a test lattice (or test grid) bearing a repeated pattern of areas (A), lines (L), and points (P) (Fig. 1). By monitoring the *chance* encounters between test lines or points and the features of interest, all the planar data necessary to reconstruct volumes, surface areas, lengths,

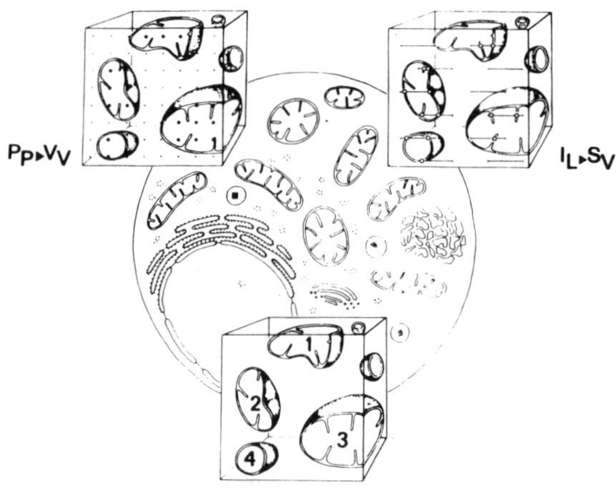

FIGURE 1. The different probes for estimation of the three main stereologic parameters: volume, surface, and number. P_p, Test point density, for estimating volume density V_v; I_L, intersection density on a test line for assessing surface density per volume S_v; N_A, numerical density of *profiles* (or transections) on a test area for estimating the numerical density of *particles* in volume N_V.

and numbers can be obtained. However, since a single section is highly unlikely to be representative of cell composition, the planar morphometric data must be taken from sets of sections (or micrographs) gathered by randomized selection procedures.

Most planar data can be obtained with relatively little effort, simply by reducing all "measurements" to the counting of discrete events, e.g., the number of lysosomal profiles (N), the number of intersections (I) between test lines and the images of the plasmalemma, or the number of test points (P) overlying profiles of mitochondria. Obviously, counting test points will not be a very precise way to estimate the profile area of a mitochondrion and counting test intersetions will not be a very precise way to estimate the boundary length of the plasmalemma. However, these measuring errors are of little practical consequence because mitochondrial area and plasmalemma boundary length will also vary from micrograph to micrograph, from tissue block to tissue block, and from one animal to the next. The reduction from three to two dimensions therefore introduces only a small increase in overall statistical uncertainty.[6,9] This important practical point has consequences for the economical design of stereological experiments, a theme taken up in Chapter 2 by Gundersen et al.

The simplicity with which planar morphometric data may be derived must be offset against the fact that they may be misleading, which affects interpretations of their biological relevance and significance. Thus, the number of nuclear transections visible per unit area of section (N_A) may not be an accurate index of the number of actual nuclei per unit volume (N_V). This follows from the fact that N_A depends on nuclear size and shape as well as on section thickness. Bigger nuclei have a greater chance of being sectioned than smaller nuclei. Moreover, N_A will tend to increase with section thickness or when nuclei become more irregular in shape.

To minimize problems of biological interpretation, it is preferable to convert planar morphometric information into stereological quantities by applying appropriate formulations.[13,14] Most stereological quantities are estimated first as ratios, otherwise referred to as "component densities."

Component Densities

These relate the specified dimensions of a microscopic component (i.e., volume V, surface area S, length L, or number N) to some suitable reference, which may or may not have the same dimensions. Because the result is a density, it is important to specify both the component and the reference. Thus, the number (N) of lysosomes (lys) in a given volume (V) of cytoplasm (cyt) is referred to as the numerical density of lysosomes in the volume of cytoplasm. For convenience, this may be abbreviated as $N(\mathrm{lys})/V(\mathrm{cyt})$ or as $N_V(\mathrm{lys, cyt})$. It may be expressed in whatever units seem most appropriate, for instance, number per cubic micrometer or per cubic centimeter.

The fundamental component densities are summarized in Table 1. Principles for their estimation are described in greater detail elsewhere.[1-16]

It is reassuring to note that reliable estimates of most of these component densities can be obtained without making any assumptions about the size, shape, connectedness, orientation, or other properties of the individual components. In general, the only precondition is that the tissue sample taken from each animal be random and independent in both orientation and position.[6,13] In the case of volume density (V_V) estimation, the condition of random and independent orientation need not be fulfilled. It must also be noted that accurate estimates of numerical densities (N_V) from random and independent sections are possible only when particle size and shape are known.[13,14] The same applies to number per reference surface (N_S), per reference length (N_L), and per reference number (N_N). Recent developments[17] have generated several methods for estimating number independent of particle size and shape.

(a) Volume Density, V_V

Reliable information about V_V can be derived for any arbitrary components appearing in randomly and independently positioned sections. The robustness of this particular component density is related to the fact that it defines the sum total of volumes of all components of a given type in a specified reference volume. This component density therefore defines aggregates rather than individuals within aggregates.

With the aid of a test lattice bearing test points, the V_V of components appearing on random independent sections (or micrographs) is obtained simply by counting test points falling on the component $P(c)$ and dividing by the total test points falling on the reference $P(r)$. Then, for all sections from all blocks from one animal,

$$\frac{V(c)}{V(r)} = \frac{P(c)}{P(r)} \quad \text{or} \quad V_V(c, r) = P_P(c, r)$$

The validity of the generic relationship $V_V = P_P$ is determined by the fact that the probability of a point placed at random in the reference falling on the

TABLE 1. Summary of Fundamental Component Densities

Component dimension	Reference dimension			
	Volume, $V(r)$	Surface, $S(r)$	Length, $L(r)^a$	Number, $N(r)$
Volume, $V(c)$	$V(c)/V(r)$			
Surface area, $S(c)$	$S(c)/V(r)$	$S(c)/S(r)$		
Length, $L(c)$	$L(c)/V(r)$	$L(c)/S(r)$	$L(c)/L(r)$	
Number, $N(c)$	$N(c)/V(r)$	$N(c)/S(r)$	$N(c)/L(r)$	$N(c)/N(r)$

^{a} Note that $L(r)$ in this table refers to the reference length of a tubular structure. In the text, under the heading surface density, $L(r)$ refers to the length of test lattice line traversing the sectioned reference.

component depends *solely* on the fraction of reference volume occupied by the component.[13]

Examples of the applications of this basic principle are offered in Chapters 4–6 by Aase, Myking, and Ree, respectively.

(b) Surface Densities, S_V, S_S, and S_L

For valid estimates of these quantities, the confrontation between sections and component surfaces must be not only random and independent in position but also isotropic (i.e., show no preferred direction of orientation in space). Two practical situations may be envisaged:

1. The component surface is isotropic so section orientation is not critical and sections of arbitrary orientation can be taken.
2. The component surface is preferentially oriented (as, for instance, is the case for the sarcolemma of skeletal muscle fibers) so section orientation must be isotropic *or* special formulas different from those below must be applied. An example of the latter is given in Chapter 8 by Mayhew, Middleton, and Ross. A new unbiased approach, requiring a special lattice to compensate for section orientation, has recently been reviewed.[18]

When the section/surface isotropy condition is met, surface densities can be estimated using a test lattice bearing points and lines. In the case of surface density S_V, one simply counts test intersections (I) between lines and component surfaces and also estimates total test line length (L) traversing the reference. The basic principle invoked is

$$\frac{S(c)}{V(r)} = 2\,\frac{I(c)}{L(r)} \qquad \text{or} \qquad S_V(c, r) = 2I_L(c, r)$$

Even the estimation of $L(r)$ may reduced to event counting,[13] a fact exploited in Chapter 7 by Halgunset.

With highly oriented tissue such as skeletal muscle, it will often be beneficial to orient section planes. For example, in transverse sections through muscle fibers the sarcolemmal surface density in muscle volume could be estimated from the special formula

$$\frac{S(f)}{V(m)} = \frac{\pi}{2}\,\frac{I(f)}{L(m)} \qquad \text{or} \qquad S_V(f, m) = \frac{\pi}{2}\,I_L(f, m)$$

where the numerical coefficient $\pi/2$ applies because of the restricted sectioning approach.

Clearly, when estimating surface densities it is crucial to appreciate the circumstances under which particular formulations apply. The same is true of length densities, which will now be considered.

(c) Length Densities, L_V, L_S, and L_L

These component densities characterize the length of some tubular or cylindrical component (e.g., capillaries in the example offered in Chapter 10) in a reference volume (L_V) on a reference surface (L_S) or along a reference length (L_L). Biological examples are also given elsewhere.[8]

Valid estimates of these densities also require randomness and independence in both position *and* orientation. Again, two practical circumstances may be envisaged:

1. The tubules are isotropic so arbitrary sections can be taken.
2. The tubules are highly orientated (as in skeletal muscle fibers) so section orientation must be isotropic *or* formulas different from that given below should be applied.

When the necessary conditions are satisfied, the number of tubule transections $Q(t)$ on the reference sectional area $A(r)$ will be proportional to their length density in a volume $L(t)/V(r)$:

$$\frac{L(t)}{V(r)} = 2\frac{Q(t)}{A(r)} \qquad \text{or} \qquad L_V(t,\, r) = 2Q_A(t,\, r)$$

where the estimation of $A(r)$ can also be reduced to event counting, as shown in Chapter 10.

Again, with highly oriented tissues it may be advantageous to restrict section planes to a preset direction. For example, in transverse sections through skeletal muscle fibers their length density in muscle volume could be estimated using

$$\frac{L(f)}{V(m)} = \frac{Q(f)}{A(m)} \qquad \text{or} \qquad L_V(f,\, m) = Q_A(f,\, m)$$

since the pertinent numerical coefficient is equal to unity in this special case.

When counting transections through tubular components, unbiased counting and selection rules must be followed.[19,20] The same applies to estimating numerical densities, which will be considered next.

(d) Numerical Densities, N_V, N_S, N_L, and N_N

The component densities described so far differ from numerical densities because they seek to define aggregates rather than each individual within an aggregate. For this reason, they are easier to estimate than numerical densities. Indeed, numerical densities cannot be estimated from random and independent sections if the particles concerned are of arbitrary size, shape, and orientation. An alternative method is described in Chapter 12 on numerical density by de Groot. Chapter 11 by Reitan describes one approach to estimating N_V that applies when size and shape can be described satisfactorily. For guidelines on the estimation of N_V in other circumstances and on methods for estimating N_S, N_L, and N_N, the reader is referred elsewhere.[8,9,13,14,21]

Attention is drawn to a recently reported method for obtaining unbiased estimates of the number of particles of any size and form in a specimen, the so-called disector method.[21] An unbiased two-dimensional frame is used for counting particles on two parallel sections separated by a distance determined by relevant dimensions of the smallest particle in the sample. The section thickness is critical and has to be measured precisely. For further details of this and other unbiased methods for estimating particle number see Sterio[21] and Gundersen.[17]

It is very important to realize that number of particles may be a redundant piece of information. Unless one is specifically interested in organellogenesis, growth (by hyperplasia versus hypertrophy), or cellular communication (e.g., by synapses or gap junctions), there may be no interpretational advantage in estimating numbers.[6,8]

Converting Densities to Absolute Quantities

Because component densities are merely ratios, meaningful interpretations of the biological importance of changes in component densities can be deduced only when they are converted into absolute volumes, surface areas, lengths, and numbers. This is commonly done by estimating the reference dimension in absolute terms. For example, if the reference happens to be the volume of an organ, this can be estimated in one or more of several different ways: by determining its shape and size, by liquid displacement, by dividing its wet weight by its specific gravity, or by cutting it into parallel slices and multiplying mean slice area by total length in the direction of sectioning.[1,3,13,22] A novel way to estimate the volumes of irregular cells from their areas of attachment to a substrate is described in Chapter 7.

CONSIDERATIONS OF SAMPLING AND STATISTICS

Before embarking on a morphometric study in which large amounts of numerical data may accumulate, it is wise to give careful forethought to the types of error that may be introduced along the way.

The realistic electron microscopist cannot hope to investigate *all* of the tissue that is available for examination under the microscope. To achieve optimal lateral resolution, thin sections covering minute fractions of the intact organ are all that can be examined. In morphometric and stereological work, it is crucial that these sections of the organ be generated by randomized sampling procedures. It is a central paradox of sampling that even randomized designs give no guarantee that the final sample will be representative of the object (e.g., organ) under investigation. However, it helps to satisfy our consciences that we have not unwittingly or deliberately selected so-called typical or interesting features appearing on the sections or micrographs.

Sampling is most efficiently performed when cutting and punching devices, as shown in Figs. 2 and 3, are used. These devices can be used when organs such as kidney or liver are perfusion-fixed and are hard enough for cutting by the equidistant razor blades. The thin, 0.8-μm sections will serve for electron microscopy and the thicker sections for light microscopy. Frozen material can be "sectioned" in a similar manner, by replacing the razor blades with sawing blades driven by a small motor. Small tissue blocks can be punched out in a systematic way by the use of a plastic template with equidistant holes (Fig. 3).

Typically, the electron microscopist might take one organ (e.g., the liver) from each of a number of animals. Later, pieces of liver will be blocked for electron microscopy and some of these blocks will provide sections from which fields of view will be selected for micrography under the microscope. This type of sampling may be described for convenience as stage sampling, since each stage of the overall selection process requires the selection of a subsample.

Stage sampling must be undertaken in such a way as to minimize two

FIGURE 2. Cutting device for preparing equidistant thin or thicker slices for electron and light microscopy, respectively.

FIGURE 3. Thin slices placed on a wet wax plate over which is placed a transparent plastic template through which a number of identically sized, equidistant holes in the form of a square have been bored. By inserting a filed-down hypodermic needle through these holes, tissue blocks are punched out. After withdrawing the hypodermic needle, a wire inserted through it is used to push the tissue block out.

main types of error: accidental error (random error) and bias (systematic error).

Minimizing Accidental Errors

The sampling design taken above for our example can be considered to be a three-stage selection process. At the *first* stage, each animal within a given experimental group is represented by one organ. Since most biological experiments are characterized by natural differences between individuals, we may denote the "biological variation" between animals by the variance S_a^2.

Because of the enforced stage sampling, this variance has other "within-animals" variances superimposed on it. Thus, in the *second* stage of selection each animal is represented by a number of tissue blocks, which also differ from one another. This variance we may denote by S_b^2. In practice, little is to be gained from taking more than one section from each block, so the *third* stage of selection may be regarded as that of choosing fields to record. These fields will also vary, with variance S_f^2.

The important statistic for comparing two groups of animals is the standard error of each group mean (SEM). In the example above, this will be represented by the square root of Os_a^2/n_a, where Os_a^2 signifies the *observed* variation between animals and n_a the total number of animals in the experimental group. The following relationships apply:

$$\text{SEM} = \sqrt{\frac{Os_a^2}{n_a}} = \sqrt{\frac{S_a^2}{n_a} + \frac{S_b^2}{n_a n_b} + \frac{S_f^2}{n_a n_b n_f}}$$

where n_b is the average number of independent blocks per animal and n_f the average number of fields sampled per block.

From this relationship it is clear that increasing n_a (the number of animals) will bring about the most effective improvement in SEM, whereas increasing n_f (the number of fields) is expected to have the least effect.

These considerations have important consequences for the optimal design of experiments, allowing for balances between estimated variances and estimated costs of sampling items (e.g., the cost of one extra animal, one extra block, or one extra field). Further details can be found elsewhere.[23-25] Methods for minimizing within-animals differences are further treated in a review.[9] These aspects of design are also considered in Chapter 2 by Gundersen et al., in Chapter 5 by Myking, and in Chapter 6 by Ree.

If each sampling stage is completed by random and independent selection processes, then the observed variance will provide a measure of the precision (reliability) of the final morphometric estimate. Precision and validity (accuracy) are synonymous only when there is no bias.

Minimizing Biases

The terms random and independent imply selection at each stage without regard to the quality or content of the items being selected. In other words, each block must have an equal and independent change of being selected, and this condition must also apply to the choice of fields. If these conditions are not satisfied, final estimates may be biased since the samples will fail to reflect accurately the characteristics of the organ from which they were drawn. The danger of this is that the validity of biological interpretations may be nullified by different degrees of bias in different experimental groups.

Generally speaking, bias can be introduced in three main ways: (1) by deficiencies in the selection processes, (2) by limitations in technique, and (3) by application of inappropriate formulas.[9]

Selection Bias

This can be avoided by careful scrutiny at each stage of the selection process.[9,13] Occasionally, bias may be introduced deliberately because it facilitates identification of cell type. An example of introducing this type of bias —which subsequently must be corrected—will be found in Chapter 4 by Aase.

Technical Bias

This may be introduced in a number of ways: tissue preparation, inadequate image resolution, insufficient specimen contrast, positive section thickness, and sectioning angle.[9,13]

Fixation, dehydration, embedding, and sectioning can alter the relative and absolute dimensions of the specimen by shrinkage, swelling, and compression. For comparative studies, technical bias may not be a serious problem provided its effects are similar in all experimental groups. In Chapter 3 by Reith et al., the influence of the type, mode, and vehicle of fixation is discussed in some detail.

Inadequate resolution of structural details on sections may also lead to technical bias, especially in systematically underestimating surface densities,[26,27] though estimates of organelle number are also vulnerable.[28] It is wise to select the minimum magnification that permits adequate resolution of the components under study since this will also minimize accidental errors.

Section thickness is also a recurrent problem because the basic principles of stereology assume that the sections are of zero thickness, i.e., true planes. Clearly, this is not the case, but the effects of this bias can be minimized by cutting sections as thinly as possible and preferably of a thickness that is less than 10 times the mean projected height of the components being analyzed.[13]

Finally, the angle at which biomembranes are sectioned can lead to bias since membranes prepared for visualization by electron microscopy tend to disappear when tilted at too large an angle within the section. This bias and methods for dealing with it are treated in Chapter 9 by Mayhew and Reith.

Estimation Bias

This usually creeps in when experimenters fail to appreciate the particular constraints under which stereological formulas can be regarded as valid. An example of problems requiring formulas different from those that apply for randomness and independence in both position and orientation is given in Chapter 8 by Mayhew et al.

THE VALUE OF MACHINE MEASUREMENT

The basic methods described above require no elaborate or expensive machinery apart from a decent computer for data handling. The lack of obvious benefit of machine measuring precision is emphasized in Chapter 2 by Gundersen et al. and in Chapter 5 by Myking. However, for progress to continue we must always keep a vigilant eye on possible future developments, a theme taken up in Chapter 13 by Rigaut.

REFERENCES

1. Aherne WA, Dunnill MS: Morphometry. London: Arnold, 1982.
2. DeHoff RT, Rhines FN: Quantitative Microscopy. New York: McGraw-Hill, 1968.
3. Dunnill MS: Quantitative methods in histology. In: Recent Advances in Clinical Pathology, series V, edited by Dyke SC, pp. 401–416. London: Churchhill, 1968.

4. Elias H, Hennig A, Schwartz DE: Stereology: applications to biomedical research. Physiol Rev 51:158–200, 1971.
5. Elias H, Hyde DM: A Guide to Practical Stereology. Basel: Karger, 1983.
6. Gundersen HJG: Stereology—or how figures for spatial shape and content are obtained by observation of structures in sections. Microsc Acta 83:409–426, 1980.
7. Loud AV, Anversa P: Biology of disease. Morphometric analysis of biologic processes. Lab Invest 50:250–261, 1984.
8. Mayhew TM: Basic stereological relationships for quantitative microscopical anatomy—a simple systematic approach. J Anat 129:95–105, 1979.
9. Mayhew TM: Stereology: progress in quantitative microscopical anatomy. In: Progress in Anatomy, volume 3, edited by Navaratnam V, Harrison RJ, pp. 81–112, Cambridge: Cambridge University Press, 1983.
10. Miles RE, Davy PJ: Precise and general conditions for the validity of a comprehensive set of stereological fundamental formulae. J Microsc 107:211–226, 1976.
11. Underwood EE: Quantitative Stereology. Reading, Mass.: Addison-Wesley, 1970.
12. Weibel ER: Stereological principles for morphometry in electron microscopic cytology. Int Rev Cytol 26:235–302, 1969.
13. Weibel ER: Stereological Methods, volume 1, Practical Methods for Biological Morphometry. London: Academic Press, 1979.
14. Weibel ER: Stereological Methods, volume 2, Theoretical Foundations. London: Academic Press, 1980.
15. Weibel ER, Bolender RP: Stereological techniques for electron microscopic morphometry. In: Principles and Techniques of Electron Microscopy, volume 3, edited by Hayat MA, pp. 237–296. New York: Van Nostrand-Reinhold, 1973.
16. Williams MA: Quantitative Methods in Biology. Oxford: North-Holland, 1977.
17. Gundersen HJG: Stereology of arbitrary particles. A review of unbiased number and size estimators and the presentation of some new ones, in memory of William R. Thompson. J Microsc 143:3–45, 1986.
18. Baddeley AJ, Gundersen HJG, Cruz-Orive LM: Estimation of surface area from vertical sections. J Microsc 142:259–276, 1985.
19. Gundersen HJG: Notes on the estimation of the numerical density of arbitrary profiles: the edge effect. J Microsc 111:219–223, 1977.
20. Miles RE: The sampling, by quadrats, of planar aggregates. J Microsc 113:257–267, 1978.
21. Sterio DC: The unbiased estimation of number and sizes of arbitrary particles using the disector. J Microsc 134:127–136, 1984.
22. Scherle W: A simple method for volumetry of organs in quantitative stereology. Mikroskopie 26:57–61, 1970.
23. Shay J: Economy of effort in electron microscope morphometry. Am J Pathol 81:503–512, 1975.
24. Gundersen HJG, Østerby R: Optimizing sampling efficiency of stereological studies in biology: or "Do more less well!" J Microsc 121:65–73, 1981.
25. Gupta M, Mayhew TM, Bedi KS, Sharma AK, White FH: Inter-animal variation and its influence on the overall precision of morphometric estimates based on nested sampling designs. J Microsc 131:147–154, 1983.
26. Gehr P, Bachofen M, Weibel ER: The normal human lung: ultrastructure and morphometric estimation of diffusion capacity. Respir Physiol 32:121–140, 1978.
27. Paumgartner D, Losa G, Weibel ER: Resolution effect on the stereological estimation of surface and volume and its interpretation in terms of fractal dimensions. J Microsc 121:51–63, 1981.
28. Mayhew TM, Williams MA: A quantitative morphological analysis of macrophage stimulation. II. Changes in granule number, size and size distributions. Cell Tissue Res 150:529–543, 1974.

part 1

experimental design

This part emphasizes the importance of considering errors at the very outset of the experimental design. Chapter 2, Gundersen et al., deals with accidental errors and their influence on the choice of experimental design and on the choice of measuring approach. For the optimal sampling design, variances at each stage must be balanced against relative costs as they apply in the particular experimental circumstances. Slavishly following previously published designs may not be the most cost-effective approach.

Chapter 3, by Reith et al., deals with systematic errors introduced by specimen preparation procedures. The influence of fixation on the preservation of cellular fine structure is a perennial problem well known to every electron microscopist. Though fixation is but one aspect of preservation and consequent biases, this chapter serves to emphasize the extent to which fixation can affect the validity of morphometric data.

chapter 2

designing efficient sampling schemes: automatic, semiautomatic, or manual image analysis for stereological studies in biology?

H. J. G. Gundersen, O. Götzsche, E. B. Jensen, and R. Østerby

INTRODUCTION

This chapter addresses the question: "Given a biological problem for the solution of which quantitative structural information is necessary, which measuring method will provide the relevant data most efficiently?" We restrict ourselves to biology because a number of practical problems and one factor of prime importance (the interindividual or biological variation) are more or less specific to biology as opposed to, for instance, materials science. "Efficiency" is used in a broad sense where we combine two requirements: the need for a *low statistical variation* in the final result(s) and the obvious necessity of getting that result with *minimal effort and cost.* (Quite often the results of measurements also have a bias, but despite the importance of such biases, which may or may not be partly correctable, they are left out in the presentation because they are usually not related to a particular measuring device.)

SAMPLING AND MEASURING PRECISION

As all biologists are well aware, meaningful quantitative information in biology is not obtained from the study of a single individual. The ever-present biological variation requires that we deal with groups and base our conclusions solely on results valid for the groups. The group of n_a individuals or animals is therefore the highest level in our sampling scheme. At the next sampling level in stereological studies we almost always have a number n_s of sections within each organ, the sections are studied with a microscope where n_f fields of vision per section are seen, and, finally, on each section we measure or estimate some structural parameter by use of an image analyzer, a semiautomatic device, or a test grid. In all cases the measuring procedure introduces or adds on a certain variation; that is, if we repeat the measurement with exactly the same field we will generally not obtain exactly the same result again. The question of how to make a sampling design in which the numbers of animals, sections, and fields are balanced against their cost is considered in a stereological context with actual examples in Ref. 1–4.

If we only had to consider the variance S_m^2 of the measuring device, the decision alluded to in the title of this chapter would seem much easier. Between two devices we would choose the one that was more precise and therefore gave, say, a 10-fold smaller measuring variance, *if* its price and the workload when using it were *less* than 10-fold larger than those of the other device. This is, of course, the argument put forward by all those who sell expensive and precise devices, and it is supported by most of the literature dealing with the question of how to measure in stereology—a literature in which the underlying assumption (that measuring precision is important) has been accepted uncritically. The argument is, however, clearly beside the point: we want to consider efficiency for the *whole experiment,* not for measuring in *a single field* of vision from a single section through a single organ.

Even if we accept the premises for the argument in favor of great measuring precisions, the implicit conclusion (that more data with greater precision from a given field are equivalent to more information) is *not* true in general.

As shown in Fig. 1, the precise measurement of, for example, the profile boundary per area, B_A, in a field of vision may be less informative than a crude point-counting estimate of intercepts between the boundary and test lines per test point hitting the area, I_p. As is well known from simple stereological equations, the two estimators $(\pi/2k)\bar{I}/\bar{P}$ and $\bar{B}/\bar{A}$ both estimate the ratio boundary per area of a structure in the field; k is the length of a test line segment. Notice, however, that the stereological relations tell us only that the *mean* values are estimators of B/A (and thereby of surface-to-volume ratios of the sectioned objects); we do not have information about *variances*. It is therefore a fact[5] that a crude and indirect estimate may contain more information than a precise and direct measure in each field because the two estimators will in general have *different variances,* and unless we know these estimator variances we do *not* know which estimator is the better one. In the example I_p has a variance that is less than two-thirds that of B_A and is therefore clearly the better estimator—evidently it is also much faster to use.

Nevertheless, it is still beside the point to take into account only the measuring precision. As outlined above, each study involves several individuals and they do not have identical values for, say, the surface of the liver mitochondrial membrane; that is, at that level of sampling there is a certain variation S_a^2 among animals. In addition, the sections from one organ will always

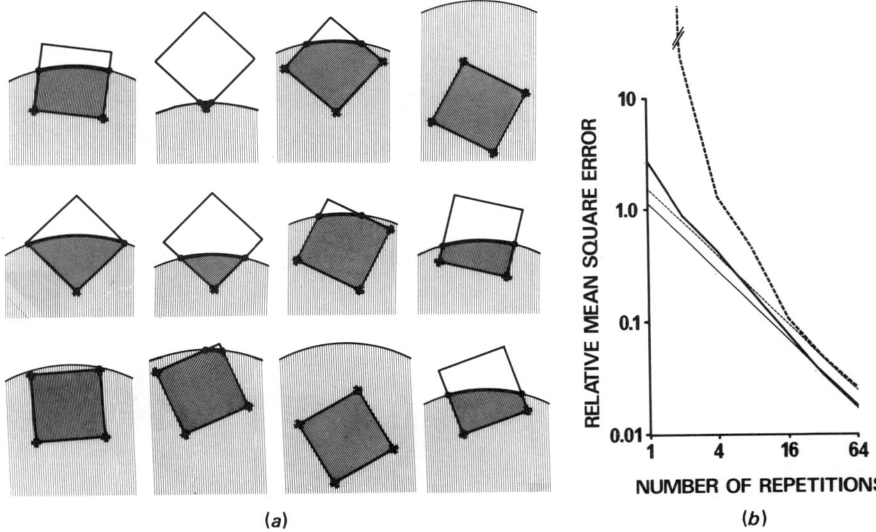

FIGURE 1. The 12 parts of (*a*) illustrate the sampling by a quadrat or a window of a large reference circle RC, the purpose being to estimate the boundary per area ratio $B(RC)/A(RC)$ of the reference circle from measurements in the sample quadrats. This can be done either by measuring in each quadrat the length B of circle boundary and the area A of circle or by counting intersection points I between circle and quadrat boundaries and number of hits P of quadrat corners in the reference circle. The two estimators $\Sigma B/\Sigma A$ and $c\Sigma I/\Sigma P$ both estimate $B(RC)/A(RC)$; c is a known constant. As shown in (*b*), the precision expressed as the relative mean square error (heavy lines) depends in both cases strongly on the number of quadrats sampled; the thin lines are theoretical predictions from large-sample theory. In all cases the variation of the point-sampled estimator (solid lines) is less than two-thirds that of the precisely measured estimator (dashed lines).

show a variation S_s^2 with respect to the parameter of interest, and so will the fields of vision within a given section. To the variation S_f^2 between fields in a section is then added the measuring variation S_m^2. We are therefore in a situation where the precision in our final stereological estimate—the standard error SE of the mean in the group—depends on several *different* types of variation in the sampling hierachy:

$$\text{SE}^2 = \frac{1}{n_a} S_a^2 + \frac{1}{n_a n_s} S_s^2 + \frac{1}{n_a n_s n_f} (S_f^2 + S_m^2) \tag{1}$$

a general formula valid under proper sampling schemes. If the organ under study shows a reproducible gradient of inhomogeneity on a set of parallel sections, an even better estimate of the sampling contribution to the final variation may be obtained by modeling the gradient, either parametrically[6] or by computer simulation.[7]

Equation (1) expresses a number of pertinent points:

1. The error variance SE^2 is inversely proportional to the number of individuals or animals in the group, n_a. (Consequence: Study as many animals as you can afford.)
2. An increase in the number of sections per organ, n_s, does *not* reduce the contribution from animal variation, S_a^2, to SE^2. (Consequence: Do not study many sections unless the organ is very inhomogeneous *and* animals are very similar.)
3. The measuring variation S_m^2 is added to the variation from field to field, S_f^2. It is therefore *never* rational to increase the measuring precision unless $S_m^2 > S_f^2$. If $S_f^2 > S_m^2$, that is, the variation from field to field is equal to or larger than the measuring variation, it is best to increase the number of *fields* studied since that will reduce the contribution from both S_f^2 and S_m^2 [see Eq. (1) again]. (Consequence: Measure crudely on a number of fields on each section with a fast method rather than precisely on a few fields with a slow device. A careful study[8] of the relative importance of S_f^2 and S_m^2 at the lowest level of sampling showed that when the time consumption is also taken into account a test grid with *few* enough points and lines is a better choice than a semiautomatic device.) Notice that when *automatic* image analyzers can handle the complex images of biological material their *speed* and not their precision will be the actual revolution. When very many fields and sections can be studied for the same overall cost in time and money there will be a reduction in the overall variance of results to the level of the variance among animals—but not, of course, below it!

The principal points outlined above reduce to a few practical rules for sampling and measuring in one animal:

1. Make approximately five sections, fewer if the organ is homogeneous, more if it is known to be very inhomogeneous *and* animals are very similar.
2. Examine 5 to 10 fields of vision per section, fewer if fields show little variation (depends on magnification among other things), more if they vary greatly *and* sections are very similar.
3. Measure on each field with precision and time consumption that are *inversely* proportional to the variation from field to field; be fast and use little precision when fields vary greatly, that is, always, unless the magnification is low.
4. Count no more than 100 to 300 points, intersections, or profiles in *one* animal and a similar number of points on the reference space. Considerably *fewer* may well be enough.[4,9]

After a pilot experiment in a new study—the analysis of which takes 1 or 2 h—one can then easily design a test system that always fulfills all of the above requirements for efficient sampling.[5,10]

The relationship mentioned above between the variation among fields and the magnification is one well known and always exploited by microscopists: The lower the magnification, the larger is the actual area seen in one field and therefore the lower the variation from one field to the next.

OPTIMIZING SAMPLING

To be more concise regarding the number of sections and fields and the measuring precision in any particular case, one must take into account the cost or time expenditure for a section or a field or the measurements. In this way it is possible to *optimize* sampling in the sense that the final variation of the results is a minimum for a given total cost of analyzing one animal. The technique is well known in statistics, and when costs and variances are known or estimated for the sampling items animals, sections, and fields one gets an estimate of the optimal number of sections and fields per animal.[1,3,4,6] The nomogram in Fig. 2 illustrates the general principle that the optimal distribution of effort in an experiment is to study a number of items in proportion to their variation and in inverse proportion to their cost.

Exactly the same principle tells us which measuring device to use: a precise, slow, and costly one if fields are very similar and relatively few need to be investigated, but a fast, inexpensive, and not precise point-counting system when fields are varying and many must be studied.

Again, sometime in the future automatic imaging machines may be able to provide the necessary speed for the analysis of *many* fields and sections. For the present, however, it is not wise to ignore the vast potential and capacity of the human eye-brain complex for dealing expediently with relatively complex images. In fact, recent techniques for the estimation of number[12] and

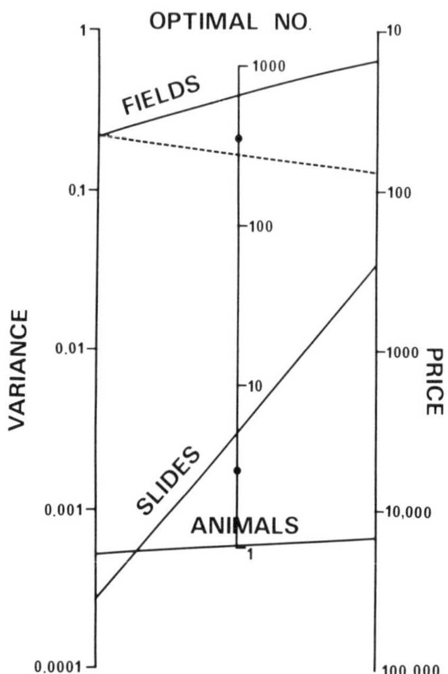

FIGURE 2. Nomogram for determining the optimal number of slides or sections and fields of vision per animal. On the left logarithmic axis the variances are entered; on the right inverse-logarithmic axis the price (money and/or time) per unit is shown. Variance and price are connected by straight lines item by item, and on the logarithmic axis in the middle the optimal number of items per animal is read off; the scale has unity at its intersection with the variance-price line for animals. The dots indicate actual numbers used in the experiment from which the example[11] is taken. The dashed variance-price line indicates how the optimal number of fields is reduced if the price per field is increased fivefold, for example, by the use of slow and expensive measuring equipment. This reduction in the number of fields studied is a crucial factor that almost always leads to *reduced* precision in the final estimate; see point 3 after Eq. (1) in the text.

mean volume[13] of particles of arbitrary shape, which effectively have broken the long-standing stereological "convexity barrier," require sampling of information for which machines are very badly suited.

CONCLUDING REMARKS

It is worth emphasizing that the principles discussed above are valid, without modification, for all stereological/morphological studies in biology. Specifically, they apply no matter what is actually measured: a mean stereological ratio of the ordinary type (V_V, S_V, L_V), an absolute dimension of a three-dimensional structure (mean cylinder or sphere diameter, membrane thickness[3]), a two-dimensionally defined profile characteristic such as shape or

Ferret diameter, or the whole distribution of some two- or three-dimensionally defined measures. In all cases there is in biology a certain variation between individuals to which is added some variation from sections through the organ, to which is again added variation from field to field, and on top of this heap of variation is added some measuring variation—but that is usually of little, if any, consequence compared to the other, inescapable variances in a biological experiment.

A properly conducted study is one where all aspects of experimental variation—not only measuring precision—are considered. The main point in doing this [Eq. (1)] is to identify the important sources of sampling variation. In the past, the number of animals has virtually always been lower than the number of sections, which has, in turn, been lower than the number of fields studied.

REFERENCES

1. Nicholson WL: Application of statistical methods in quantitative microscopy. J Microsc 113:223–239, 1978.
2. Gundersen HJG: Stereology—or how figures for spatial shape and content are obtained by observation of structures in sections. Microsc Acta 83:409–420, 1980.
3. Gundersen HJG, Østerby R: Optimizing sampling efficiency of stereological studies in biology or "Do more less well!" J Microsc 121:65–73, 1981.
4. Gundersen HJG, Østerby R: Sampling efficiency and biological variation in stereology. Mikroskopie 37(Suppl):143–148, 1980.
5. Jensen EB, Gundersen HJG: Stereological ratio estimation based on counts from integral test systems. J Microsc 125:161–166, 1982.
6. Kroustrup JP, Gundersen HJG: Sampling problems in a heterogeneous organ: quantitation of relative and total volume of pancreatic islets by light microscopy. J Microsc 132:43–55, 1983.
7. Cruz-Orive LM, Myking OA: Stereological estimation of volume ratios by systematic sections. J Microsc 122:143–157, 1980.
8. Mathieu O, Cruz-Orive LM, Hoppeler H, Weibel ER: Measuring error and sampling variation in stereology: comparison of the efficiency of various methods for planar image analysis. J Microsc 121:75–88, 1981.
9. Gundersen HJG, Boysen M, Reith A: Digitizer-tablet or point counting in biomorphometry? Proc 3d Eur Symp Stereol, Stereol Iugosl 3(Suppl 1):205–210, 1981.
10. Gundersen HJG: Stereology and sampling of biological surfaces. In: The Analysis of Organic and Biological Surfaces, edited by Echlin P, pp. 478–508. New York: Wiley, 1984.
11. Gundersen HJG, Götzsche O, Østerby R: Sampling efficiency in morphometry simplified. Metab Bone Dis Related Res 25:281–289, 1980.
12. Sterio DC: The unbiased estimation of number and sizes of arbitrary particles using the disector. J Microsc 134:127–136, 1984.
13. Gundersen HJG, Jensen EB: Stereological estimation of the mean volume of arbitrary particles observed on random sections. J. Microsc 138:127–142, 1985.

chapter 3

importance of fixation for the cellular and subcellular tissue composition using rat liver as an example

Albrecht Reith, Michel Kraemer, and Jeannie Vassy

INTRODUCTION

Specimen preparation is a crucial part of a morphometric investigation since the influences of fixation, dehydration, and embedding on shape and size are well known.[1-3] This chapter will deal with the influence of fixation, which is the most varied step in preparation procedures, whereas dehydration and embedding follow standard steps. A detailed discussion of the influences of dehydration and embedding on cell structure may be found in the recent book by Revel et al.[4]

The importance of fixatives, modes of fixation, and buffers has been shown for many organs under various conditions.[5-8] Studies on liver have demonstrated considerable discrepancies among fixatives in the preservation of fine structure.[6,9-11] The study described here was performed to investigate in an objective way the influence of fixation by use of the same liver and identical preparative steps. For this purpose we studied the most commonly used:

1. Fixatives: glutaraldehyde, paraformaldehyde, and osmium.
2. Modes of fixation: perfusion versus immersion.
3. Vehicles for fixatives: cacodylate and phosphate buffer.

Rat liver was used as a model and mitochondria were used to judge the preservation of fine structure, as it is known that these organelles are very

Laboratory for Electron Microscopy and Morphometry, Department of Pathology, Norwegian Radium Hospital and Cancer Institute and Norwegian Cancer Society, Oslo 3, Norway

Laboratoire de Biologie du Développement, U.E.R. Biomedicale, Université Paris XIII, F-93012, Bobigny-Cedex, France sensitive to fixatives.[6,10,11] The conclusions drawn from this study are, however, not restricted to liver but are valid for other organs as well.

MATERIAL AND METHODS

Experimental Groups

For convenience two experimental groups were studied:

First experimental group. The left lobe of three male Wistar rats weighing 250, 252, and 250 g was excised and immersion-fixed and the remainder of the liver was perfusion-fixed. Five fixation types were chosen for study, and the purpose was to

1. Change the mode of fixation (experiments 1 and 2).
2. Change the type of fixative, but keep the same buffer (experiment 1 versus experiment 4 and experiment 3 versus experiment 5).
3. Change the type of fixative vehicle but keep the same fixative (experiment 2 versus experiment 3 and experiment 4 versus experiment 5).

Second experimental group. The left lobe of the liver of three male Wistar rats weighing 250 g was used for immersion fixation with different concentrations of paraformaldehyde (experiments 6 to 9). The details of the fixation procedures are summarized in Table 1.

Histological Methods

The left liver lobe was sliced and 1-mm^3 blocks were immersed for 2 h at room temperature for glutaraldehyde or at 4°C for osmium. For paraformaldehyde immersion fixation the blocks were fixed for 10 or 4 h at 4°C.

For perfusion fixation, the rest of the liver was perfused through the venae portae for 3 to 5 minutes at a pressure of 80 mm Hg after opening the vena cava. Tissue blocks (1 mm^3) were cut and further fixed for 2 hours followed by postosmication for 1 h. All tissue pieces were dehydrated (10-min steps of 50, 70, 80, and 100% ethanol and 100% propylene), embedded (Epon), and polymerized under absolutely identical conditions in one continuous operation.

Sampling

Five blocks per rat were selected by lottery. For light microscopy 1-μm-thick toluidine blue-stained sections were used. Five micrographs per block were taken at ×250 and enlarged to ×1000 together with a lattice test system (spacing $d = 10$ μm) for analysis.

Electron Microscopy

Silver to gray sections were stained with lead citrate and examined at 60 kV in a Philips EM 300. Five micrographs were made at a magnification of ×5000 and were enlarged together with a lattice test system (spacing $d = 2$

TABLE 1. Composition of the Fixative Solutions[a]

Experiment	Fixation	Washing	Postfixation	Washing
1. Perfusion	1% Glutaraldehyde + 0.1 M cacodylate buffer, 275 mosm	0.1 M cacodylate buffer + 0.1 M sucrose, 290 mosm, 24 h at 4°C	1% OsO_4 + 0.1 M cacodylate buffer + 0.1 M sucrose, 325 mosm, 1 h at 4°C	0.1 M cacodylate buffer + 0.1 M sucrose, 290 mosm, 1 h at 4°C
2. Immersion	1% Glutaraldehyde + 0.1 M cacodylate, 275 mosm, 2 h at 20°C	0.1 M cacodylate buffer + 0.1 M sucrose, 290 mosm, 24 h at 4°C	1% OsO_4 + 0.1 M cacodylate buffer + 0.1 M sucrose, 325 mosm, 1 h at 4°C	0.1 M cacodylate buffer + 0.1 M sucrose, 290 mosm, 1 h at 4°C
3. Immersion	1% Glutaraldehyde + 0.135 M phosphate, 390 mosm, 2 h at 20°C	0.135 M phosphate buffer + 0.1 M sucrose, 365 mosm, 24 h at 4°C	1% OsO_4 + 0.135 M phosphate buffer, 340 mosm, 1 h at 4°C	0.135 M phosphate buffer + 0.1 M sucrose, 365 mosm, 1 h at 4°C
4. Immersion	1% OsO_4 + 0.1 M cacodylate buffer, 215 mosm, 2 h at 4°C	0.1 M cacodylate buffer + 0.05 M sucrose, 235 mosm, 1 h at 4°C		
5. Immersion	1% OsO_4 + 0.135 M phosphate buffer, 340 mosm, 2 h at 4°C	0.135 M phosphate buffer + 0.1 M sucrose, 365 mosm, 1 h at 4°C		
6. Immersion	10% Paraformaldehyde + 0.1 M phosphate buffer, 2000 osm, 10 h at 4°C			
7. Immersion	6% Paraformaldehyde + 0.25% glutaraldehyde + 0.1 M phosphate buffer, 2000 osm, 10 h at 4°C	0.1 M phosphate buffer, 230 nosm, 24 h at 4°C	1% OsO_4 + 0.1 M phosphate, 280 mosm, 1 h at 4°C	0.1 M phosphate buffer, 230 mosm, 1 h at 4°C
8. Immersion	4% Paraformaldehyde + 0.1 M phosphate buffer, 1415 mosm, 4 h at 4°C			
9. Immersion	1% Paraformaldehyde + 0.1 M phosphate buffer, 560 mosm, 4 h at 4 °C			

[a] The osmolarities are the final measurements; pH of all solutions was 7.4.

μm) to a final magnification of × 15,000. The test line length for the surface density estimation was 210 μm.

Stereological Parameters

The components estimated included the volume densities of the

- Extrahepatocytic space (VVEX), that is, Disse space, sinusoids, bile canaliculi, portal and terminal veins, and all nonhepatocyte cells.
- Hepatocytes (VVH).
- Hepatocyte nuclei (VVNH).
- Hepatocyte cytoplasm (VVCYT).
- Hepatocyte mitochondria (VVMIT, CYT).

Moreover, the surface density of the outer membrane (SVMOCYT), the surface-to-volume density of the chondroime (SVMO/VVMIT, CYT) and the mean profile area of mitochondria (am) were measured.

Student's t-test was used for statistical analysis; p values of 0.05 or less were reported as statistically significant.

RESULTS

Qualitative Results

Perfusion and Immersion Fixation with Glutaraldehyde and Osmium

The perfusion-fixed material (experiment 1) was homogeneous throughout the block (Fig. 1). In immersion fixation a zonation with respect to cellular and subcellular structure was observed. The depth of the zone of good preservation varied from three to several cell layers. From there inward, dark and light cells were the prominent features of fixation disturbances at the cellular level. At the electron microscopic level, extended cisternae of rough endoplasmic reticulum (RER) and perinuclear spaces, swollen mitochondria, and distorted, sometimes "exploded," mitochondria were observed.

For cacodylate-buffered glutaraldehyde (experiment 2) one or two cell layers were well preserved, but from 60 μm below the block's cut surface the fine structure became disturbed (Fig. 2).

For phosphate-buffered glutaraldehyde (experiment 3) the zone of good preservation was three cell layers (Fig. 3).

For osmium-fixed material (experiments 4 and 5) the results were similar, the zone of good preservation being about three or four cell layers (Fig. 4).

Immersion Fixation with Various Paraformaldehyde Concentrations in Phosphate Buffer

With 10% paraformaldehyde (experiment 6) the first six cell layers were quite well preserved. From there (~150 μm) inward, organelle-free cytoplasmic areas were observed beside clusters of mitochondria and RER (Fig. 5).

For 6% paraformaldehyde with 0.25 glutaraldehyde (experiment 7) the first three cell layers were homogeneous and well preserved. Light and dark cells occurred 80 μm below the cut surface, that is, in the fourth and fifth cell layers. Light cells often displayed larger organelle-free cytoplasmic areas (Fig. 6).

For 4% paraformaldehyde the hepatocytes of the second and third layers displayed wrinkled nuclei, a slightly extended RER, and dark mitochondria. In the fourth to sixth cell layers the RER perinuclear space became conspicuously extended and the mitochondria were darker than in the preceding layer (Fig. 7). From there inward to about 90 μm below the cut edge, the RER was less extended but organelle-free cytoplasmic areas were observed with adjacent clusters of RER and mitochondria.

For 1% paraformaldehyde the liver cells in the first to sixth layers were homogeneous. The cells in the seventh and eighth layers (i.e., 90–120 μm) were somewhat darker than in the preceding layers (Fig. 8). From 120 μm inward the hepatocytes displayed fixation changes such as extended RER and swelling of mitochondria with an electron-light interior.

Quantitative Changes

The quantitative evaluation showed significant differences between the perfusion- and immersion-fixed material. There were also significant differences between the various immersion experiments with regard to two fixatives or two vehicles used. The morphological changes in the different experiments, described in detail in Table 1 are shown as percentage differences relative to glutaraldehyde perfusion (= 100%) in Fig. 9.

At the cellular level the volume densities of hepatocytes (VVH, L) and liver cell cytoplasm (VVH/CYT, L) increased by 20% when comparing immersion to perfusion fixation, whereas their nuclei (VVHN, L) increased less. However, for paraformaldehyde there was a decrease of the volumetric density of nuclei, whereas the cytoplasm increased.

At the subcellular level the volume density of mitochondria (VVMCYT) showed large differences. Compared to perfusion fixation, glutaraldehyde-fixed mitochondria increased by one-third, for osmium fixation there was no change, whereas for paraformaldehyde there was a conspicuous decrease.

The surface density of the outer membrane of mitochondria (SVMOCYT), compared to perfusion fixation, decreased slightly in glutaraldehyde- or osmium-fixed material (except for phosphate-buffered glutaraldehyde material) in spite of the increase in the volume density of mitochondria. For paraformaldehyde there was a still more pronounced decrease (70%) of outer surface density.

The changes were most conspicuously expressed in the assessment of the

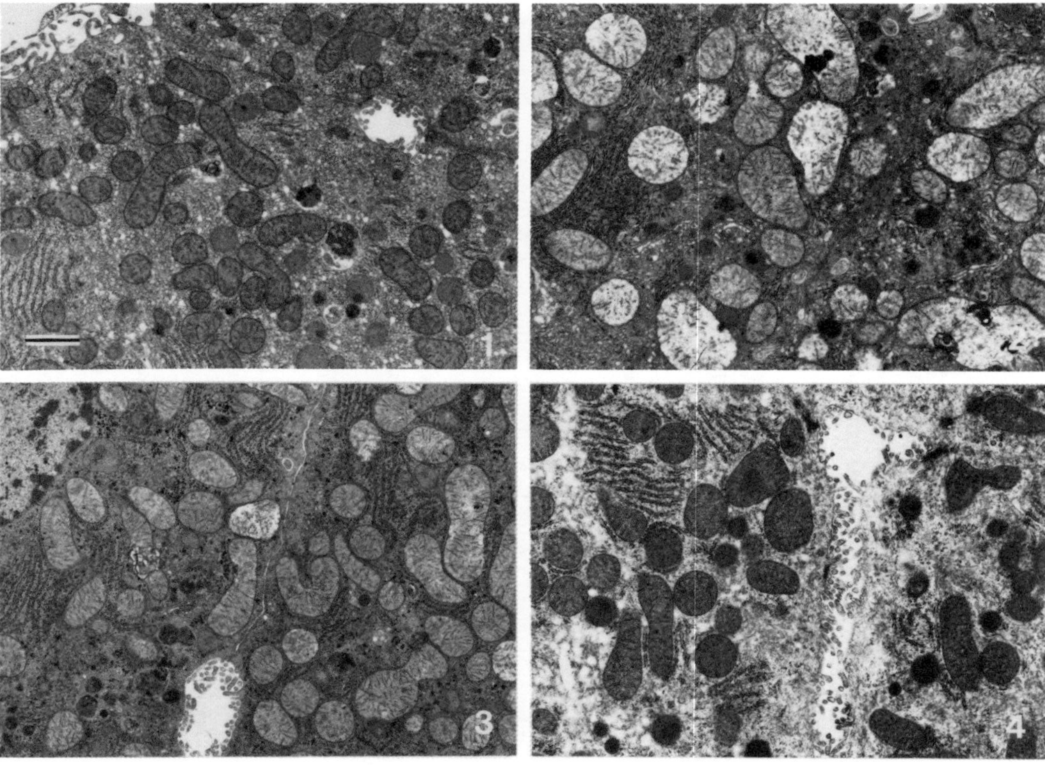

FIGURES 1–4. Influence of perfusion and immersion fixation with glutaraldehyde and osmium in cacodylate and phosphate buffer on the fine structure of rat liver (scale bar, 1 μm). Fig. 1: 1% glutaraldehyde in 0.1 M cacodylate buffer, perfusion fixed. Fig. 2: Same as Fig. 1, but immersion fixed. For one or two cell layers below the cut edge there was good preservation (from 60 μm below the surface the fine structure became disturbed). Fig. 3: Immersion fixation with phosphate-buffered 1% glutaraldehyde. Fig. 4: Immersion fixation with 1% OsO_4 in 0.1 M cacodylate buffer in the zone of good preservation (three or four cell layers below cut edge.)

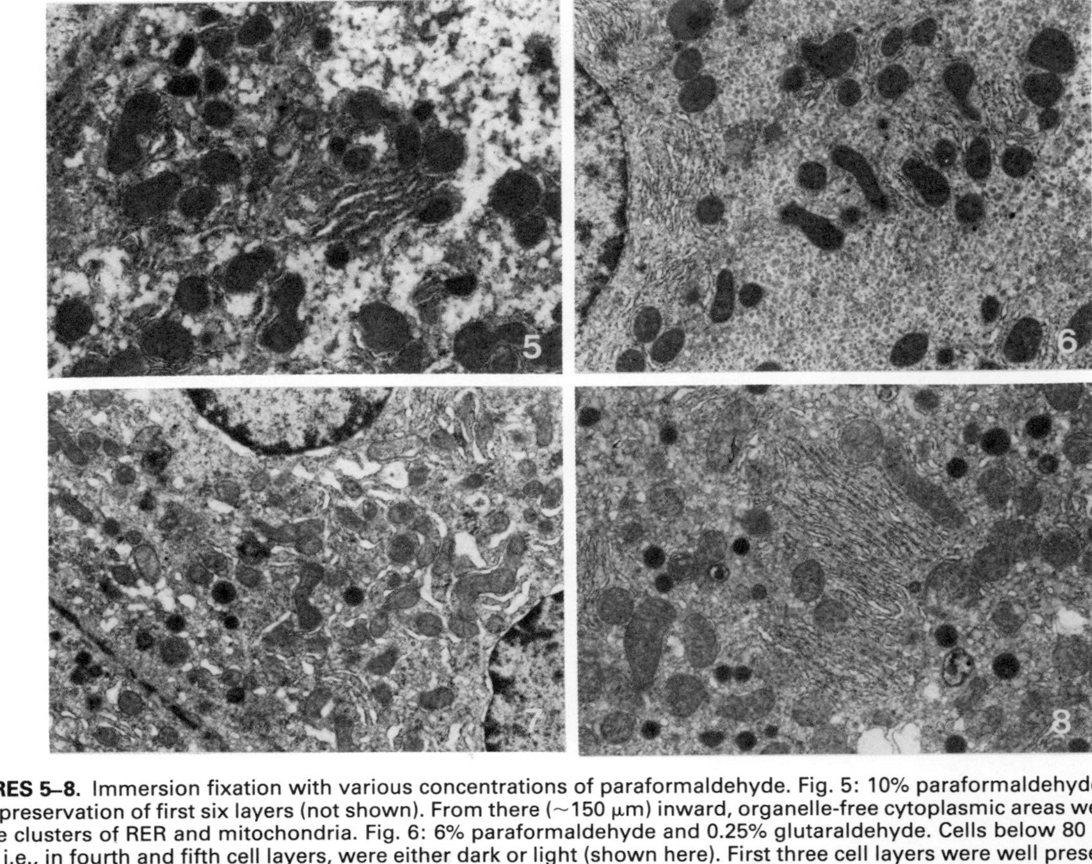

FIGURES 5–8. Immersion fixation with various concentrations of paraformaldehyde. Fig. 5: 10% paraformaldehyde. There was good preservation of first six layers (not shown). From there ($\sim$150 μm) inward, organelle-free cytoplasmic areas were observed beside clusters of RER and mitochondria. Fig. 6: 6% paraformaldehyde and 0.25% glutaraldehyde. Cells below 80 μm from cut edge, i.e., in fourth and fifth cell layers, were either dark or light (shown here). First three cell layers were well preserved. Fig. 7: 4% paraformaldehyde. Cells in second and third layers displayed wrinkled nuclei, slightly extended RER, and dark mitochondria. At fourth to sixth layers (shown here), RER and perinuclear space became conspicuously extended and mitochondria darker than in preceding layers. Fig. 8: Cells in first to sixth layers were homogeneous. Thereafter (90–120 μm, seventh, eighth, or ninth layers) they were somewhat darker (shown here). Below 120 μm fine structure was disrupted.

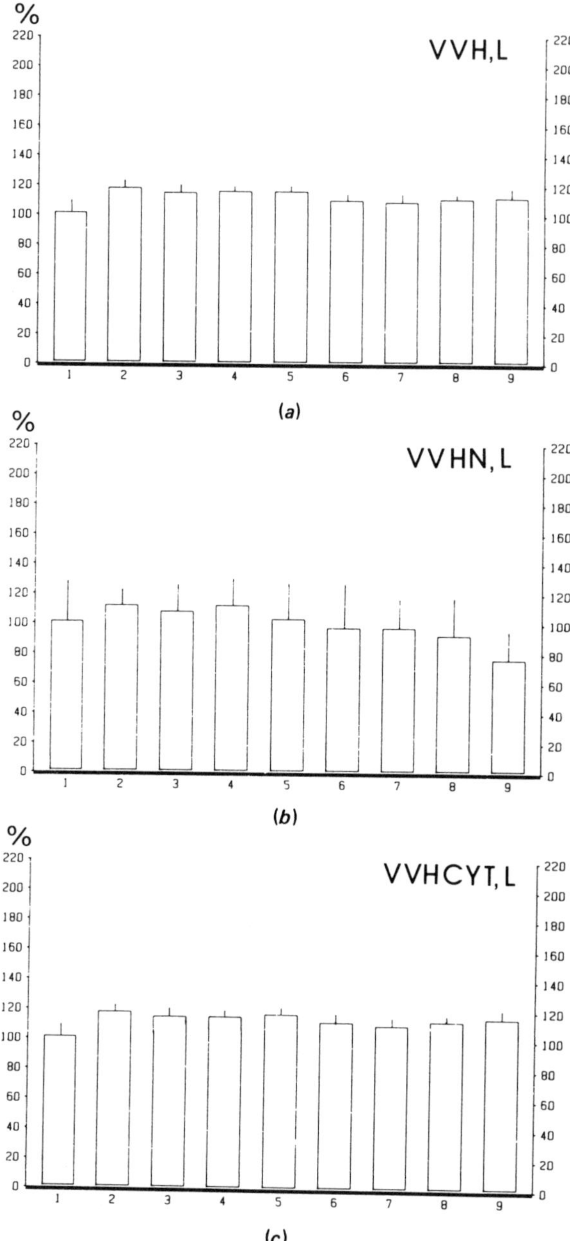

FIGURE 9. Quantitative changes in the fixation experiments (1–9) are shown as percentage differences relative to glutaraldehyde perfusion fixation (= 100%). For each experiment the mean standard deviation is shown. (a–c) Light microscopy and (d–f) electron microscopy results. To see the influence of fixation, note particularly the difference in mean area of mitochondrial profiles; the mitochondrial size is doubled in immersion compared to perfusion fixation, even though the same fixa-

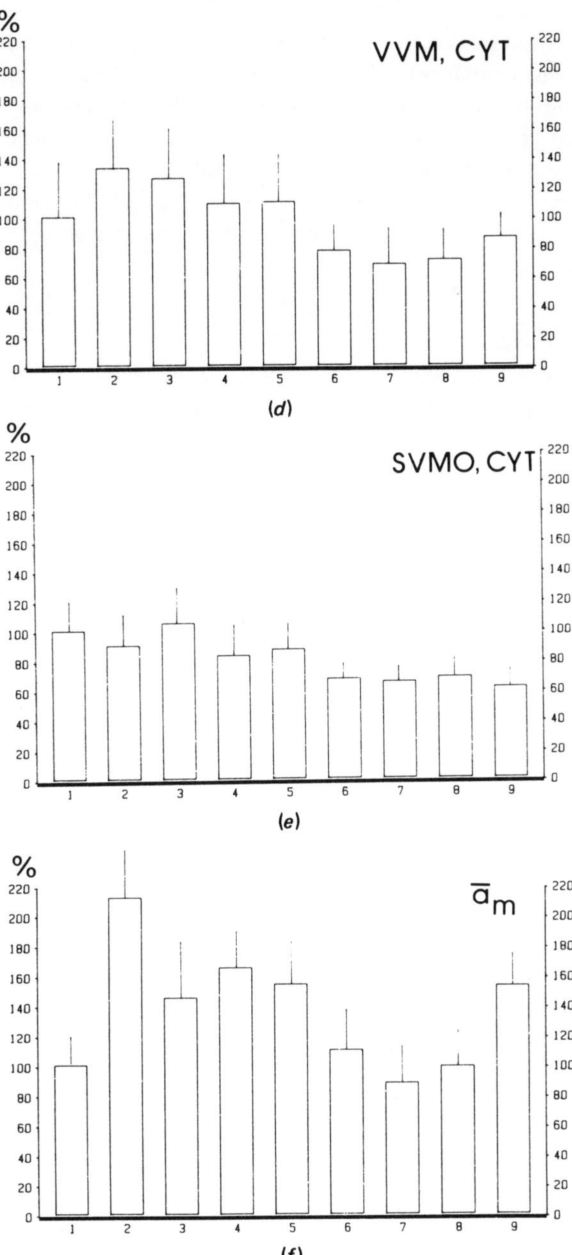

tion and buffer were applied. Experimental conditions are given in Table 1. (*a*) Volume density of hepatocytes (VVH, L); (*b*) volume density of hepatocyte nuclei (VVHN, L); (*c*) volume density of hepatocyte cytoplasm (VVHCYT, L); (*d*) volume density of mitochondria (VVM, CYT); (*e*) surface density of outer membrane (SVMO, CYT); (*f*) mean profile area $\bar{a}_m$.

profile area of mitochondria (a), which was increased 1.5- to 2-fold in glutar-aldehyde- and osmium-fixed material. However, for paraformaldehyde the mean profile area of mitochondria remained relatively unchanged compared to perfusion. The one exception was the 1% paraformaldehyde experiment, in which the profile area increased by one-half.

More detailed descriptions of the results of changing the mode of fixation, type of fixative, and buffers can be found in Kraemer et al.[12] and Reith et al.[13]

DISCUSSION

The study clearly demonstrates the influence of fixation on the cellular and subcellular organization of liver cells. Four main conclusions can be drawn from this investigation:

1. Fixation by perfusion is superior to immersion.
2. In immersion fixation one has to take into account that a zonation occurs in the blocks with respect to the quality of fine-structure preservation. The layer with well-preserved cells can be rather small.
3. Large differences in the quantitative assessment of the mitochondrial compartment can occur, depending on the fixation procedure used.
4. Assessment of the *profile size* of mitochondria easily reveals fixation influences in morphometric/stereologic work.

That zonation occurs within immersion-fixed specimen blocks is common knowledge to electron microscopists and was also demonstrated in earlier studies on fixation.[14] That the zones of good preservation may be rather small depending on the fixative type or fixative buffer used is, however, not so well known. Qualitative changes such as gross distortions of the fine structure, extended RER and perinuclear spaces, swollen mitochondria with electron microscopically light interior, or disrupted cristae are easily detected by the observer. However, quantitative changes that precede the qualitative changes are easily overlooked if they are not very dramatic. Therefore it may be advisable to use flat-embedded material and to section in such a way that micrographs of the sections can always be taken at the same distance from the block's cut surface. It will not always be easy to predict the dimensions of the zone with good preservation when using different tissues. Our numbers for liver tissue may be a rough estimate. When extrapolating from liver tissue one must take into account that immersion fixation of the liver is facilitated by the large well-developed intraparenchymal spaces in this tissue, mainly the sinusoids and the space of Disse. In studies by Rømert and Matthiessen[10] taking preservation of mitochondria as an indicator, it was observed that with 2.5% glutaraldehyde and 0.1 M cacodylate buffer, the zone was within a range of 100 μm from the block edge. Use of 0.8% glutaraldehyde narrowed this zone to 50 μm. Our result of 60 μm with 1% glutaraldehyde and 0.1 M cacodylate buffer (experiment 2) lies in between these two values. The buffer type is also important, as shown by Fahimi,[15] Mathieu et al.,[16] Rømert and Mattheissen,[10] and in our experiment 3 when changing to phosphate buffer, which resulted in a 90-μm zone. The superior quality of the phosphate buffer (close to 300 mosm) in glutaraldehyde immersion fixation compared to cacodylate buffer (200 mosm) may also be related to the osmotic pressure of the vehicles. According to Brunk et al.[5] the buffer should be isosmolar, that is, 300 mosm in fixation, and they claimed that the glutaraldehyde does not contribute to the *effective* osmotic pressure of the fixative. This is probably true for immersion fixation but may be less important, as our experiment 1 shows, in perfusion fixation, where the distance of fixation by diffusion is only about 10 μm compared to large diffusion distances from the block's cut edge into its interior.

Paraformaldehyde clearly contributes to the *effective* osmotic pressure, as shown by the effect of the different concentrations of paraformaldehyde on the mitochondria size. There was a 20% larger mitochondria fraction and a 40% larger profile area of mitochondria in 1% than in 10% paraformaldehyde.

For immersion fixation the penetration rate of a fixative is crucial. According to Palade,[17] the penetration rate of OsO_4, is 1 mm/h. The penetration follows the general formula of Hopwood,[18] $d = K\sqrt{t}$, where d is the penetration distance, t the fixation time, and K a constant for the fixative used. In the case of OsO_4, $K = 0.2$. Glutaraldehyde is still slow ($K = 0.34$), whereas paraformaldehyde is rather fast with $K = 2.0$. To make use of the good fixation qualities of glutaraldehyde and the fast fixation by paraformaldehyde, a mixture of both aldehydes has been applied successfully.[19-22]

CONCLUSION

In quantitative work, the choice of fixative, mode of fixation, and fixative buffer should be made with particular care. It may be advisable for sensitive material, that is, pathologically altered material, to perform a series of trial experiments in which the tonicity of the fixative solution is changed. For a good example see Mathieu et al.[16] In work on bioptic material of the nasal mucosa, 3% glutaraldehyde in 0.1 M cacodylate buffer with 0.1 M sucrose has given good preservation for light and electron microscopic material analyzed quantitatively.[23]

Fixation

If possible, perfusion fixation is always the first choice. The effort put into this part of the experiment pays dividends in time gained at the later "sampling" step; that is, no time need be spent in searching for sections of well-preserved tissue. One can also expect less variation between blocks, so it may be possible to reduce the number of blocks to be analyzed.

If immersion fixation is performed, the combination of glutaraldehyde and paraformaldehyde[22] may also be a good choice.

The tonicity of the buffer should be slightly more than 300 mosm. The tonicity of the OsO_4 solution is again important but not as crucial.

Care should also be taken that the washing buffer before osmication has this tonicity[3] (e.g., replacing the aldehyde by sucrose).

Zonal Variation in Preservation

Consideration must be given to the zones of good and bad preservation in specimens fixed by immersion. It may be advisable to take micrograph samples only at a certain distance from the block cut edge by making use of flat-embedded material. If possible, it is always preferable to use perfusion fixation, which minimizes the diffusion distances of the fixative solution to about 10 μm. Homogeneity of the preservation of fine structure is the reward for the more laborious work at the beginning of a stereologic study.

Dehydration/Embedding

Finally, it must be mentioned that not only fixation but also dehydration and embedding procedures influence the quantitative composition of the cell structure. Their importance in electron microscopy is dealt with in the recent book by Revel et al.[4]

REFERENCES

1. Reith A, Barnard T, Rohr HP: Stereology of cellular reaction patterns. CRC Crit Rev Toxicol 4:219–269, 1976.
2. Weibel ER: Stereological Methods, volume 1, Practical Methods for Biological Morphometry. London: Academic Press, 1979.
3. Williams MA: Autoradiography and immunocytochemistry. In: Practical Methods in Electron Microscopy, volume 6, edited by Glauert, AM, pp. 1–217. Amsterdam: North-Holland, 1977.
4. Revel J-P, Barnard T, Haggis GH (eds.): The Science of Biological Specimen Preparation for Microscopy and Microanalysis. AMF O'Hare, Ill.: Scanning Electron Microscopy Inc., 1984.
5. Brunk U, Collins VP, Arro E: The fixation, dehydration, drying and coating of cultured cells for SEM. J Microsc (Oxf) 123:121–131, 1980.
6. Loud AV, Barany WC, Pack BA: Quantitative evaluation of cytoplasmic structures in electron micrographs. Lab Invest 16:258–270, 1985.
7. Maunsbach AB: The influence of different fixatives and fixation methods on the ultrastructure of rat kidney proximate tubule cells. II. Effects of varying osmolality, ionic strength, buffer system and fixation concentration of glutaraldehyde solutions. J Ultrastruct Res 15:283–309, 1966.
8. Eins S, Wilhelms E: Assessment of preparative volume changes in the central nervous tissue using automatic image analysis. Microscope 24:29–38, 1976.
9. Bahr GF, Bloom G, Friberg U: Volume changes of tissues in physiological fluids during fixation in OsO_4 or formaldehyde and during subsequent treatment. Exp Cell Res 12:342–355, 1957.
10. Rømert P, Matthiessen ME: Swelling of mitochondria in immersion-fixed liver tissue. Effect of various fixatives and of delayed fixation. Acta Anat 109:332–338, 1981

11. Stempack J, Laurencin M: Mitochondrial form in hepatic parenchymal cells in rats of several ages. Am J Anat 145:261–281, 1976.
12. Kraemer M, Vassy J, Chalumeau MT, Reith A: Influence de la fixation au paraformaldéhyde sur la préservation qualitative du quelques paramètres stéréologiques du foie de rat adulte. Arch Anat Microsc Morphol Exp 72:279–298, 1983.
13. Reith A, Kraemer M, Vassy J, Chalumeau MT: The influence of mode of fixation, type of fixative and vehicles on the same rat liver. A stereologic/morphometric study by light and electron microscopy. Scanning Electron Microsc II:645–651, 1984.
14. Hopwood D: Some aspects of fixation with glutaraldehyde. A biochemical and histological comparison of the effects of formaldehyde and glutaraldehyde fixation on various enzymes and glycogen, with a note on penetration of glutaraldehyde into liver. J Anat 101:83–93, 1967.
15. Fahimi HD: Perfusion and immersion fixation of rat liver with glutaraldehyde. Lab Invest 16:736–750, 1967.
16. Mathieu O, Claasen H, Weibel ER: Differential effect of glutaraldehyde and buffer osmolarity on cell dimensions. A study on lung tissue. J Ultrastruct Res 63:20–34, 1978.
17. Palade GE: A study of fixation for electron microscopy. J Exp Med 95:285–297, 1952.
18. Hopwood D: Theoretical and practical aspects of glutaraldehyde fixation. Histochem J 4:267–303, 1972.
19. Chambers RW, Bowling MC, Grimley PM: Glutaraldehyde fixation in routine histopathology. Arch Pathol 85:18–30, 1968.
20. Ericsson JLE, Biberfeld P: Studies on aldehyde fixation. Fixation rates and their relation to fine structure and some histochemical reactions in liver. Lab Invest 17:281–289, 1968.
21. Karnovsky MJ: A formaldehyde-glutaraldehyde fixative of high osmolarity for use in electron microscopy. J Cell Biol 17:137A–138A, 1965.
22. McDowell EM, Trump BF: Histologic fixatives suitable for diagnostic light and electron microscopy. Arch Pathol Lab Med 100:405–414, 1976.
23. Boysen M, Reith A: Stereological analysis of nasal mucosa. III. Stepwise alterations in cellular and subcellular components of pseudostratified, metaplastic and dysplastic epithelium in nickel workers. Virchows Arch B Cell Pathol 40:311–325, 1982.

part 2

volume measurements

This part comprises three chapters dealing with volume densities (V_v) of tissue components that are sometimes difficult to recognize or occur in very small quantities and for which special sampling strategies must be developed.

Chapter 4, by Aase, deals with the problem of sampling only cells that are sectioned through the nucleus (so-called nuclear-biased sampling) and taking into account all cell profiles in order to arrive at reliable absolute and relative volumes of cell nuclei.

In Chapter 5 Myking addresses the important question of how to sample volume densities in an efficient way. Different sampling strategies are compared. Myking also takes up the often-asked question: ''Is it preferable to use electronic tracing devices or the 'stone-age method' of point counting?''

Ree's Chapter 6 describes an efficient method for analyzing the volume composition of rarely occurring components by combining light microscopy and electron microscopy.

Since an efficient sampling strategy effectively cuts the workload in morphometric/stereologic studies, these chapters may be particularly interesting for newcomers to the field of quantitation in morphological studies.

chapter 4

nuclear-biased sampling in epithelia: studies on human gastric parietal cells

Steinar Aase

INTRODUCTION

Different sampling methods have been used in stereological studies of gastric parietal cells. In previous studies at the light microscopic and electron microscopic levels, we analyzed all parietal cells seen in our sections.[1,2] Other investigators have selected for morphometry only parietal cell profiles containing certain components. This is called component-biased sampling. Zalewsky and Moody selected only profiles containing both secretory membrane and cell nucleus.[3] In a study of liver parenchymal cells, Loud et al. analyzed only cell profiles containing the nucleus.[4] This nuclear-biased sampling method has also been employed in most morphometric investigations of gastric parietal cells.[5–10] In stereology, component-biased sampling may, when uncorrected, lead to overestimation of the relative volume occupied by the component in question. Nuclear-biased studies of parietal cells in animals[3,6,7,9] have given a volume density of nuclei up to fourfold larger than that obtained in our investigations of human tissue.[1,2] The work described here was a stereological study of human parietal cells in 1-μm-thick Epon-embedded sections. It was performed to compare the nuclear volume densities resulting from unbiased and nuclear-biased sampling. In particular, it was desired to test the applicability of the Konwiński-Kozlowski formula,[11] originally developed to correct for a biased volume density of nuclei in lymphocytes.

MATERIALS AND METHODS

Gastric biopsies were taken with a hydraulic biopsy capsule from seven healthy volunteers 18–34 years of age. Simultaneous visual control with a thin gastroscope was used to ensure that all biopsies were taken about 10 cm distal to the cardia, at the transition between the greater curvature and the posterior wall of the stomach. The tissue samples were prepared for electron microscopy, the results of which have been published.[2] All biopsies used for the present study of cell nuclei were taken after an overnight fast, during basal (nonstimulated) conditions. Fixation was carried out for 2 h at 4°C in 3.5% glutaraldehyde, followed by rinsing in buffer and fixation for 2 h in 1% OsO_4. The fixatives were made up in 0.1 M sodium phosphate buffer, pH 7.4, containing 0.05 M NaCl. After fixation, the tissue samples were dehydrated in a graded series of ethanol, passed through propylene oxide, infiltrated, and finally embedded in Epon. After polymerization for 3 days at 60°C, 1-μm-thick sections were cut with an LKB ultramicrotome. The sections were stained with methylene blue, azure II, and basic fuchsin.[12] For light microscopic morphometry the sections were examined with a Leitz research microscope, model Orthoplan, using an ×63 Leitz planapochromatic oil immersion objective with a numerical aperture of 1.40. In one section from each of the seven subjects, 30 pictures were photographed in a systematic preestablished manner.[2] The negatives were enlarged in the copying process, giving paper prints with a final magnification of ×1380 for the morphometric analysis (Fig. 1).

For the morphometry we used point-counting methods.[13] The pictures were covered one by one with a transparent plastic sheet showing a morphometric square lattice test system.[13] In this test system, parallel lines were spaced 1 cm apart, and the intersections between horizontal and vertical lines were used as test points. The position and orientation of the lattice were random and independent in each micrograph. We counted all test points (P_i) lying over nuclei of parietal cells. In addition, we counted all points (P_r) lying over parietal cell profiles that contained a visible nucleus. The volume density (V_{vo}) of nuclei in parietal cells was calculated according to the formula $V_{vo} = P_i/P_r$.[13] Since parietal cell profiles without visible nuclei had been neglected, the calculated volume density was an overestimation of the true value. This overestimation was reduced by use of the formula of Konwiński and Kozlowski,[11] who stated that

$$V_{vt} = \left(\frac{3V_{vo}}{V_{vo} + 2} \right)^{3/2}$$

where V_{vt} is the true volume density and V_{vo} is the volume density observed experimentally, using a nuclear-biased sampling method. The formula was originally developed for spherical lymphocytes, each having a single spherical and centrally located nucleus.[11]

The nuclear volume densities calculated with the formula of Konwiński and Kozlowski were compared to those obtained in a study using an unbiased sampling method. The results of this unbiased study have been published.[2]

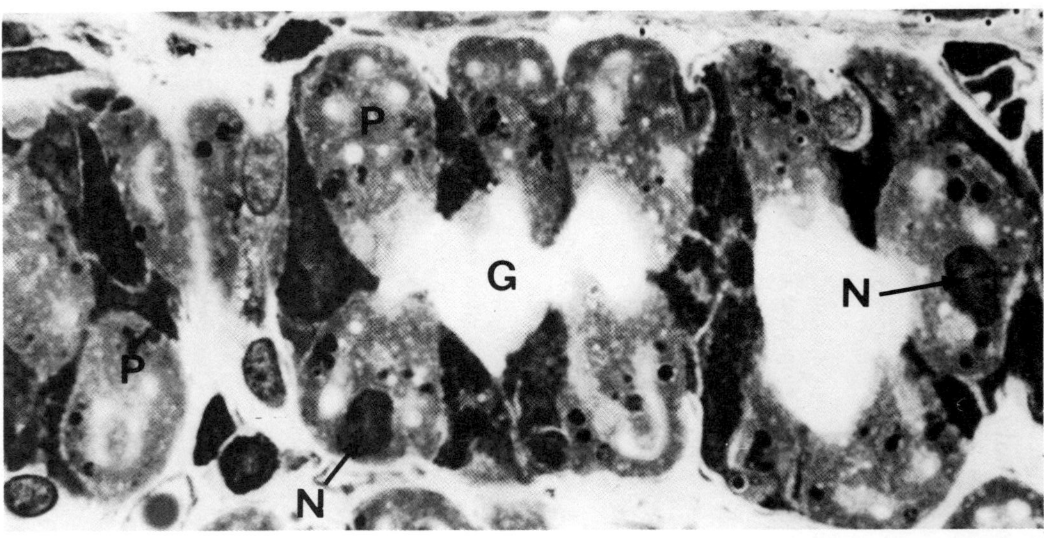

FIGURE 1. Section through a tubular gland of the gastric mucosa. G, Glandular lumen; P, parietal cell; N, nucleus of parietal cell. The 1-μm-thick section was stained with methylene blue, azure II, and basic fuchsin. Picture taken through a ×63 Leitz planapochromatic oil immersion objective with a numerical aperture of 1.40. Final magnification ×1380 (same as that used for the morphometric registrations).

We used the same square lattice test system described above and the same pictures. Now we counted all points (P_i) lying over nuclei of parietal cells, as described above, and also counted the total number of points (P_t) lying over all parietal cell profiles, regardless of whether the nucleus was detectable in the profiles. The volume density of nuclei was calculated from the formula $V_v = P_i/P_t$. We also counted the total number (N_i) of nuclei, using the unbiased counting rule of Gundersen.[14] The average absolute volume ($\overline{V}_i$) of the parietal cell nucleus was then calculated from

$$\overline{V}_i = \frac{\beta}{k}\left(A_p \frac{\Sigma P_i}{\Sigma N_i}\right)^{3/2}$$

This equation was derived from a formula originally developed by Weibel and Gomez[15] to calculate the numerical density[16] of particles in a volume. Here A_p is the area corresponding to one point in the test system and β and k are constants, β depending on shape and k on the size distribution of the particles.[15,16] Details of this method are published elsewhere.[2]

RESULTS

Most of the parietal cell profiles that were seen in the 1-μm-thick sections did not contain the nucleus. This point is illustrated in Fig. 1.

Table 1 presents the calculated absolute volume of the parietal cell nucleus in each of the seven subjects. The median size of the nucleus was 170 μm^3 (range 127–206 μm^3). Table 1 also gives the volume density of the nuclei. Based on analysis of all parietal cells in the pictures, we found a median value of 5.6% for the volume density of the nuclei (range 3.7–7.5%). From registrations of only cells containing nuclear profiles, we calculated a median

TABLE 1. Results of Stereological Studies on the Nuclei of Parietal Cells in Seven Healthy Subjects[a]

Subject	N_{nucl}	$\overline{V}_{nucl}$ (μ^3)	$V_{v(nucl,cell)1}$ (%)	$V_{v(nucl,cell)2}$ (%)	$V_{v(nucl,cell)2corr}$ (%)
1	316	185	5.3	11.1	6.3
2	180	189	4.9	11.1	6.3
3	250	145	5.6	11.3	6.4
4	312	135	3.7	10.8	6.0
5	193	127	6.6	13.9	8.6
6	168	206	5.9	12.7	7.6
7	184	170	7.5	14.6	9.2
Median	193	170	5.6	11.3	6.4

[a] N_{nucl}, Number of nuclei counted; $\overline{V}_{nucl}$, average absolute volume of nucleus; $V_{v(nucl,cell)1}$, volume density of nuclei in parietal cells, based on analysis of all such cells in the pictures; $V_{v(nucl,cell)2}$, volume density of nuclei in parietal cells, based on analysis of only cell profiles containing a nucleus; $V_{v(nucl,cell)2corr}$, $V_{v(nucl,cell)2}$ corrected by use of the formula of Konwiński and Kozlowski.

volume density of 11.3% (range 10.8–14.6%). The values found with the last method mentioned were corrected by use of the formula of Konwiński and Kozlowski. We then obtained a median volume density of 6.4% (range 6.0–9.2%), which is close to the results achieved with the unbiased sampling method.

DISCUSSION

Loud's sampling method was originally used for electron microscopic studies of liver parenchymal cells.[4] In the present work we used the method for light microscopic morphometry on the nucleus of the parietal cell. This is valid since we used Epon sections that were thin compared to the diameter of the nucleus.

If the nucleus constitutes a very large fraction of a cell, most sections through that cell will also hit the nucleus. This is true even if the sections are thin. When the nucleus is small compared to the size of the cell, many random sections may be cut through the cell without slicing the nucleus, provided the sections are thin compared to the nuclear diameter. In this respect, it is important to note that most parietal cell profiles in our 1-μm-thick sections did not contain the nucleus (Fig. 1). If this majority of cells were excluded from the stereological analysis, one would considerably overestimate the nuclear volume density if the systematic sampling error was not compensated for.

In our studies on unbiased samples at the electron microscopic level,[1,2] the values for the nuclear volume density were found to vary considerably between individuals. This is to be expected, since relatively few nuclei were analyzed. Yet the average value for a group of individuals was about the same as that presently reported from light microscopy. This nuclear volume density of about 6% is lower than the values found in electron microscopic studies of animals, using a nuclear-biased sampling method: Zalewsky and Moody[3] found a value of about 10% in dogs. Helander[7] found a nuclear volume density of about 15% in parietal cells of normal unfasted rats. Blom and Helander[9] reported that the nucleus constitutes up to 20% of the parietal cell under various experimental conditions in the rat. Most of their volume densities were not corrected for nuclear-biased sampling errors. There has been some uncertainty about the use of the formula of Konwiński and Kozlowski for such corrections, since the formula was originally developed for spherical lymphocytes in suspension. Blom and Helander[9] stated that the parietal cells often resemble truncated cones and that it is therefore doubtful whether an adequate correction can be achieved.

The shape of the parietal cells varies considerably. Some of this variation is shown in Fig. 1. The present work, however, shows that the formula of Konwiński and Kozlowski gives a rather good correction, at least for human parietal cells. The median volume density of the nuclei found with this formula was 6.4%, as compared with 5.6% using the unbiased sampling method.

We consider it important to correct for systematic sampling errors wherever possible. The overestimation of the nuclear volume density resulting from nuclear-biased sampling is corrected rather well by use of the Konwiński-Kozlowski formula, as demonstrated here. This sampling method, however, also introduces errors into the calculation of the volume densities of all cytoplasmic components, and at present there are no methods to compensate for these errors when the organelles are unevenly distributed in the cytoplasm. Simultaneously with the overestimation of the nuclear volume density, the volume densities of cytoplasmic organelles are underestimated when they are expressed as fractions of the total cell volume. If the organelles are unevenly distributed, the underestimation is largest for those found mostly in peripheral parts of the cell. Therefore, an unbiased sampling method may be even more important in electron microscopic investigations of organelles than in light microscopic studies of cell nuclei. In nuclear-biased studies at the electron microscopic level, the relative volumes of the organelles should be given not as fractions of the whole cell volume but as fractions of the cytoplasmic volume. By using the cytoplasm as the reference space one obtains the correct fractional volumes in spite of the nuclear-biased sampling, provided the organelles in question are evenly distributed in the cytoplasm.

Furthermore, in some electron microscopic studies of gastric parietal cells, only those cell profiles were sampled in which the nucleus was judged to have a certain magnitude, given as 3–25% of the profile area in one study.[8] Since this was based on a qualitative judgment, it introduced some subjectivity into the sampling. We consider it important to avoid subjective criteria in our electron microscopic studies of parietal cells.[1,2]

Other investigators have pointed out the dangers of nuclear-biased sampling of other cell types. Mayhew and Cruz discussed the problem of component-biased sampling of macrophages, liver cells, and neurons.[17] They also derived correction formulas for stereological values resulting from biased samples.[17–19] Mayhew and Cruz pointed out that component-biased sampling is still a useful and accepted sampling approach when cell identification is critical.[17] The gastric parietal cells, however, have a highly characteristic ultrastructure. The most characteristic features are the tubulovesicles and the secretory canaliculus, in addition to the high content of mitochondria.[2,6] In our electron microscopic studies, the presence or absence of a nuclear profile did not contribute significantly to the identification of these cells.[1,2] For this reason, nuclear-biased sampling seems to be unwarranted. In other cell types, such as plasma cells and megakaryocytes, the nucleus has a more characteristic structure. In these cases cell identification may depend more critically on sections passing through the nucleus. Biased sampling should be used only in cases where it is necessary for identification purposes. Today, we often find it difficult to tell whether there are species differences in the ultrastructural composition of the parietal cells or whether the differences in the results of different investigators are due to sampling errors. This has been discussed in connection with our stereological study of parietal cells in healthy subjects.[2] By avoiding all unnecessary bias and subjectivity, one increases the biological significance of the results and makes it easier to compare the morphometric values published by different investigators.

REFERENCES

1. Aase S, Roland M: Light and electron microscopical studies of parietal cells before and one year after proximal gastric vagotomy in duodenal ulcer patients. Scand J Gastroenterol 12:417–420, 1977.
2. Aase S, Dahl E, Roland M, Hars R: Morphometric studies of parietal cells during basal conditions and during stimulation with pentagastrin in healthy subjects. Scand J Gastroenterol 18:913–923, 1983.
3. Zalewsky CA, Moody FG: Stereological analysis of the parietal cell during acid secretion and inhibition. Gastroenterology 73:66–74, 1977.
4. Loud AV, Barany WC, Pack BA: Quantitative evaluation of cytoplasmic structures in electron micrographs. Lab Invest 14:996–1008, 1965.
5. Frexinos J, Carballido M, Louis A, Ribet A: Effects of pentagastrin stimulation on human parietal cells. Am J Dig Dis 16:1065–1074, 1971.
6. Helander HF, Hirschowitz BI: Quantitative ultrastructural studies on gastric parietal cells. Gastroenterology 63:951–961, 1972.
7. Helander HF: Stereological changes in rat parietal cells after vagotomy and antrectomy. Gastroenterology 71:1010–1018, 1976.
8. Ivey KJ, Tarnawski A, Sherman D, Krause WJ, Ackman K, Burks M, Hewett J: Quantitative ultrastructural analysis of the human parietal cell during acid inhibition and increase of gastric potential differences by glucagon. Gut 21:3–8, 1980.
9. Blom H, Helander HF: Quantitative ultrastructural studies on parietal cell regeneration in experimental ulcers in rat gastric mucosa. Gastroenterology 80:334–343, 1981.
10. Black JA, Forte TM, Forte JG: Inhibition of HCl secretion and the effects on ultrastructure and electrical resistance in isolated piglet gastric mucosa. Gastroenterology 81:509–519, 1981.
11. Konwiński M, Kozlowski T: Morphometric study of normal and phytohemagglutinin-stimulated lymphocytes. Z Zellforsch Mikrosk Anat 129:500–507, 1972.
12. Humphrey CD, Pittman FE: A simple methylene blue–azure II–basic fuchsin stain for epoxy-embedded tissue sections. Stain Technol 49:9–14, 1974.
13. Weibel ER: Stereological principles for morphometry in electron microscopic cytology. Int Rev Cytol 26:235–302, 1969.
14. Gundersen HJG: Stereology—or how figures for spatial shape and content are obtained by observation of structures in sections. Microsc Acta 83:409–426, 1980.
15. Weibel ER, Gomez DM: A principle for counting tissue structures on random sections. J Appl Physiol 17:343–348, 1962.
16. Weibel ER: Stereological Methods, volume 2, pp. 140–152. London: Academic Press, 1980.
17. Mayhew TM, Cruz LM: Stereological correction procedures for estimating true volume proportions from biased samples. J Microsc 99:287–299, 1973.
18. Mayhew TM, Cruz Orive LM: Some stereological correction formulae with particular applications in quantitative neurohistology. J Neurol Sci 26:503–509, 1975.
19. Cruz Orive LM: Correction of stereological parameters from biased samples on nucleated particle phases, I. Nuclear volume fractions. J Microsc 106:1–18, 1976.

chapter 5

studies on the volumetric composition of lymph nodes: problems of efficient sampling and the use of point counting versus digitizer tablets

Andreas O. Myking

INTRODUCTION

Lymph nodes, like the lymphatic system in general, have for the past 30 years been extensively studied and the achievements have accumulated in a very large literature. Against this background, the limited interest in lymph node stereology is difficult to explain. Since the mid-1960s[1] it has been known that lymph nodes consist of functionally different compartments (Fig. 1). Interpretations of the shape and size of lymph node compartments were previously based mainly on traditional histology, with a few based on semi-quantitative measurements as an adjuvant method.[2,3] However, the need to clarify important aspects of lymph node morphology and function has in recent years led to more extensive studies of shape[4,5] and volume.[6–8]

The search for practical methods of measuring the volume or weight of lymph node compartments has been motivated mainly by the need for exact data in order to evaluate the total proliferative capacity of the subunits. It has been shown that, under chronic stimulation with a contact sensitizer, the prolonged increased proliferative capacity of the paracortex is dependent solely on the increase of its volume.[9] Interest in the paracortex in this context is also due to the fact that cellular changes may be inadequate indicators of its activity, and volume determination appears to be a safer parameter.[10]

The author thanks the editors of this book for constructive review of the manuscript. Dr. L. M. Cruz-Orive contributed valuably by suggesting improvements to an earlier version of the chapter.

GENERAL PROBLEMS IN VOLUME ESTIMATION FOR LYMPH NODE COMPARTMENTS

Volumetric studies are usually based on volume *fraction* analysis. Provided the reference volume can be derived for the organ as a whole, the volumes of its different components are easily calculated when their volume fractions are known. For lymph nodes a convenient estimate of volume or size is based on their weights. The weights of the different compartments are then given by their volume fractions, assuming that the specific weights of the different compartments are equal.

Volume fraction analysis in general is based on the Delesse principle, which defines the volume ratio of a given component as being equal to the fraction of the area (in a section) occupied by that component (for a review see Weibel[11]). For components with a randomized and homogeneous spatial distribution, measurements in one section will give a fairly accurate volume

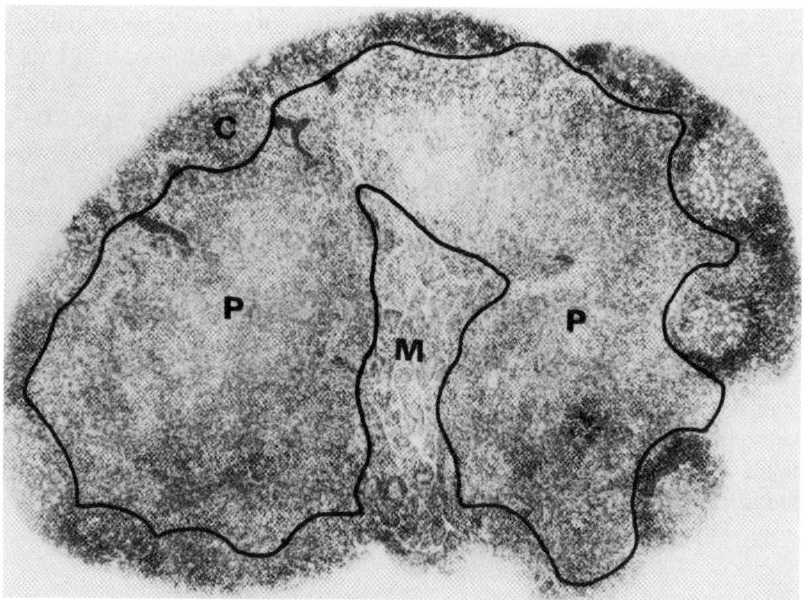

FIGURE 1. Section of mouse inguinal lymph node showing the topography and relationship between the different lymph node compartments. The cortical zone (C) is composed of densely packed lymphocytes and contains several prominent germinal centers. In the paracortex (P) the cellular density is somewhat less pronounced. The medulla (M) occupies the most central parts. The regular distribution of the different compartments is characteristic of inguinal lymph nodes in mice and inspired the development of the star specimen model (see text). Epon-embedded material, toluidine blue, ×24. (Reproduced from Ref. 6, p. 128, by kind permission of the editor.)

ratio estimate. When the components are not randomized and homogeneous, as in lymph nodes (Fig. 1), a sampling problem must be solved regarding the number and positions of the sections required.

In conclusion, volume estimation for lymph node compartments includes two main steps: weighing of the total lymph node and volume fraction analysis of its different subunits.

WEIGHING OF LYMPH NODES

This point requires consideration because the total lymph node weight is the basis for calculation of the weights of the lymph node compartments. It is obvious that weighing such a small organ involves problems, partly because it may be difficult to separate the node from adjoining tissue and partly because the weight of the fresh lymph node may be influenced significantly by factors such as room temperature, humidity, and the time from removal to weighing. It is also important that any standardized processing for weighing does not impair the quality of the histology.

In our work the nodes were fixed directly after removal and thereafter dehydrated in 100% ethanol for at least 3 days. After dehydration they were carefully dissected free from adjoining tissue. At this stage the cleaning process was easy to perform, with a much better demarcation between the capsule and adjoining tissue than in the unfixed state. By this procedure, an approximation of dry weights was obtained with a weight reduction from the fresh state of approximately 40%. Thereafter the weights remained stable, with individual variations not exceeding the general weighing error (Fig. 2). After dehydration, the time to weighing is therefore not critical. Before being weighed, the nodes were placed on filter paper at room temperature for ½ min. The procedure did not influence the results of further processing for light microscopic examination.

VOLUME FRACTION ANALYSIS

Volume or volume fraction analysis may be based on serial sectioning, which is in principle independent of the shape, size, and spatial distribution of the structures involved. This also holds for lymph nodes, for which hitherto only problems regarding volume analysis of the paracortex have been examined. The volume fraction of the paracortex v/V may be estimated from $\sum_1^{n_2} a / \sum_1^{n_1} A$, where a denotes the area of the paracortex in a single section and A the total lymph node area in the same section (Fig. 3). When point counting is used as an alternative procedure, a and A are replaced respectively by p and P, where p is the number of points hitting the paracortex and P the number of points falling within the total lymph node boundaries. In the summations, n_1 and n_2 denote the number of sections required for estimating the total lymph node volume and the volume of the paracortex, respectively.

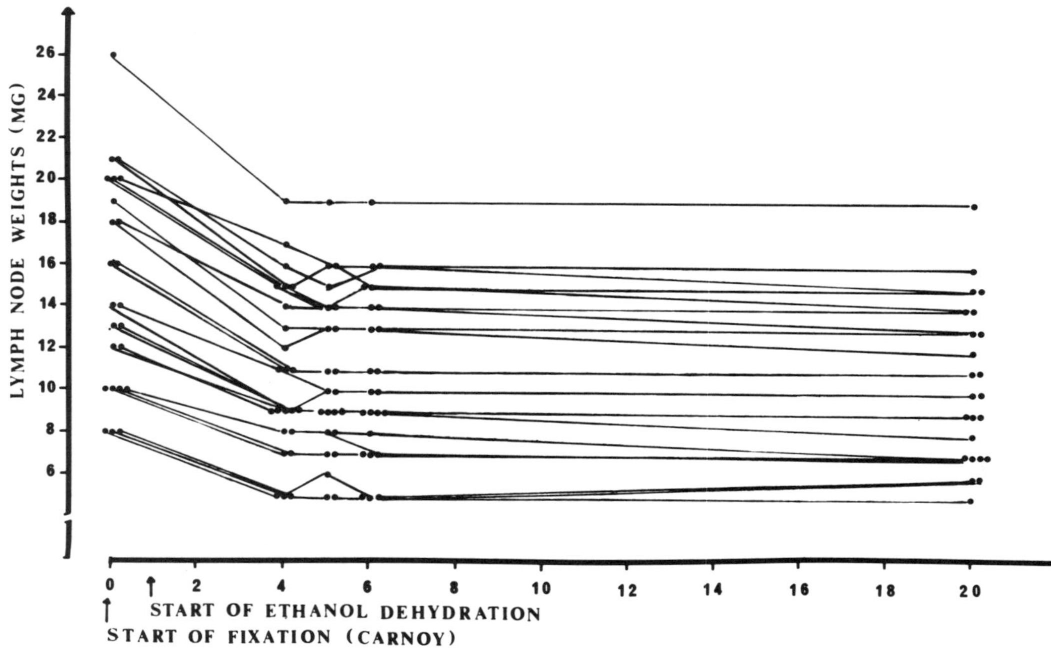

FIGURE 2. Relationship between lymph node weights and duration of dehydration in concentrated ethanol. Each point represents one lymph node. There is nearly a 40% reduction in weight during the first 3 days of dehydration. Thereafter no further weight decrease is seen.

When n is increased to infinity, the calculated volume fraction equals the true volume fraction, provided that a and A are measured with sufficient accuracy. An adaptation of this principle by taking systematic step sections at a distance of 120 μm has been used to calculate "true volume proportions," to which the results of other procedures can be compared. Obviously, even this adapatation (systematic step sections) of the ideal principle is too time-consuming for practical use. This motivates efforts to find solutions better suited for practical work.

Single Sections

Figures 4 and 5 illustrate the relationship between the position of sections and the paracortical volume fraction. In each case, positions of sections may be found where the volume proportion (area proportion) of the paracortex in a specific section equals the true paracortical volume proportion. Such positions are given by the points of intersection between the line for true volume proportions and the curve illustrating the relationship between positions and volume proportions. Although considerable variation exists, the graphs are

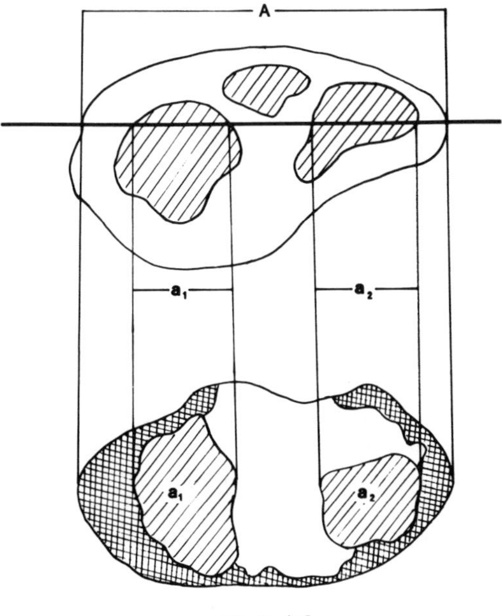

FIGURE 3. Sectioning of a guinea pig lymph node; A denotes the total lymph node section area and $a_{1,2}$ the section area of the the paracortex.

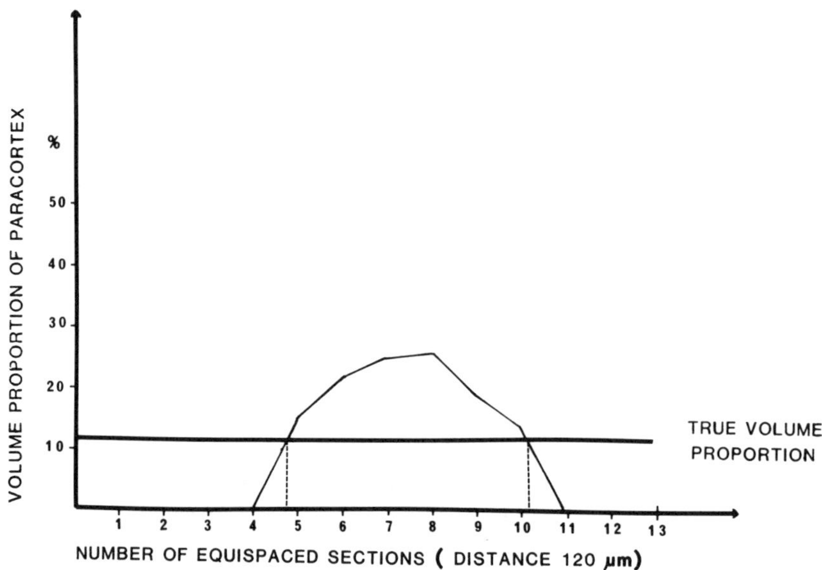

FIGURE 4. Relationship between the area proportion (volume proportion) of the paracortex and the positions of equispaced step sections of a mouse lymph node. Note that centrally placed sections will be considerably overrepresented in relation to the true volume proportion. The dashed lines represent positions of sections where the area fraction of the paracortex equals the true volume proportion.

typical for mice and guinea pig lymph nodes and express a more regular spatial distribution and a more central location of the paracortex in mice. In the graph for the mouse lymph node two such positions exist (between sections 4 and 5 and approximately at section 10). In the guinea pig graph, there are four points of intersection (these nodes are considerably larger, so the probability of hitting a "true position" appears to be approximately the same for the two species).

For practical purposes, the use of a single section as a basis for volume fraction analysis of the paracortex would be most convenient. However, because there is considerable variation within individuals, between individuals, and between different species, the "true positions" as illustrated in Figs. 4 and 5 would require serial sectioning to be defined and therefore this is not a practical procedure. To seek a practical solution, the problem was investigated in two different ways: in mice by the creation of a geometric model, and in mice and guinea pigs by the use of a single centrally located section.

The "Star Specimen" Model

Taking advantage of specimen shape, it is often possible with the aid of a geometric model to estimate volume fraction from a single section with a defined position. A requisite in most situations is precise orientation of the specimen before sectioning. From serial sectioning and reconstruction of mouse inguinal lymph nodes, a close similarity between the paracortex in the node and a star specimen model became apparent. Theoretical considerations and a more extensive description have been presented.[6] The essential point is that an arbitrary section through the "star point" gives the volume ratio of the inner body to the outer from the corresponding area fraction to the power 1.5. The results were tested against estimates from serial sections of 11 *inguinal* mouse lymph nodes, randomly selected from days 0, 4, 9, and 12 after repeated stimulation with the contact sensitizing agent oxazolone. With such stimulation, the paracortex increases considerably in size. The nodes were cut serially at 6 μm in a plane perpendicular to the top-hilus axis, and the individual star points were determined. The position of the star points turned out to be fairly constant at one-third of the top-hilus caliper length. The estimates based on the star sections came very close to the "true volume proportions" in individual nodes.[6] Later, the model was tested extensively in a large series of stimulated *auricular* mouse and guinea pig lymph nodes, and barely acceptable agreement was obtained for individual nodes (L. M. Cruz-Orive and A. O. Myking, unpublished results). For groups of lymph nodes, however, there was a constant difference between the star specimen estimates and the true values (Fig. 6), and the model may therefore be used to follow

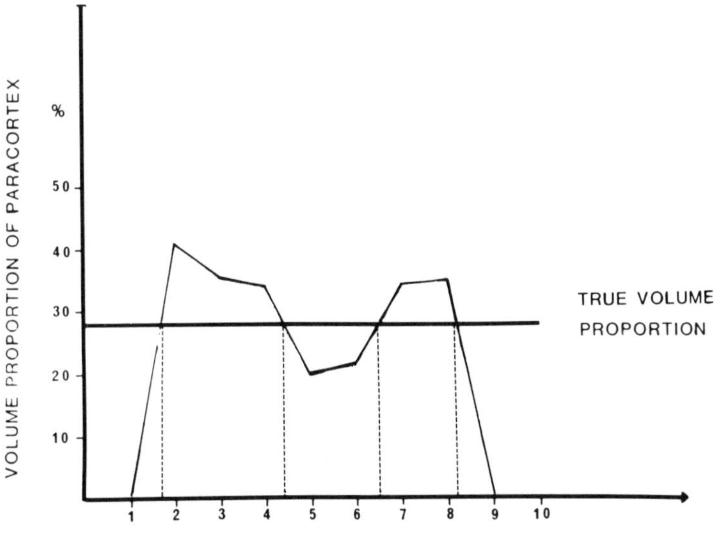

FIGURE 5. Relationship between the area proportion (volume proportion) of the paracortex and the position of equispaced step sections of a guinea pig lymph node. Note that centrally placed sections will be underrepresented in relation to the true paracortical volume proportion. The dashed lines represent positions of sections where the area fraction of the paracortex equals the true volume proportion.

alterations in groups of lymph nodes. This principle is equally valid for the use of midzonal single sections and will therefore be treated more extensively in the next chapter. For *practical work,* the star specimen may be used for mouse *inguinal* lymph nodes. However, other procedures appear to be more attractive.

Midzonal Sections

The midzonal section has been defined as the largest section when the total lymph node is cut serially.[8] Considering the usual bean shape of a lymph node, this is in nearly every instance a centrally placed section. No further premises need to be given, so the plane of sectioning may be varied at random. In fact, two situations have been tested, one where the lymph nodes were strictly orientated so that the plane of sectioning was perpendicular to the top-hilus axis, the other where the plane of cutting was random. No difference was found in the results. The model has also been shown to be fairly robust against deviations from the midzonal position, as sections 120

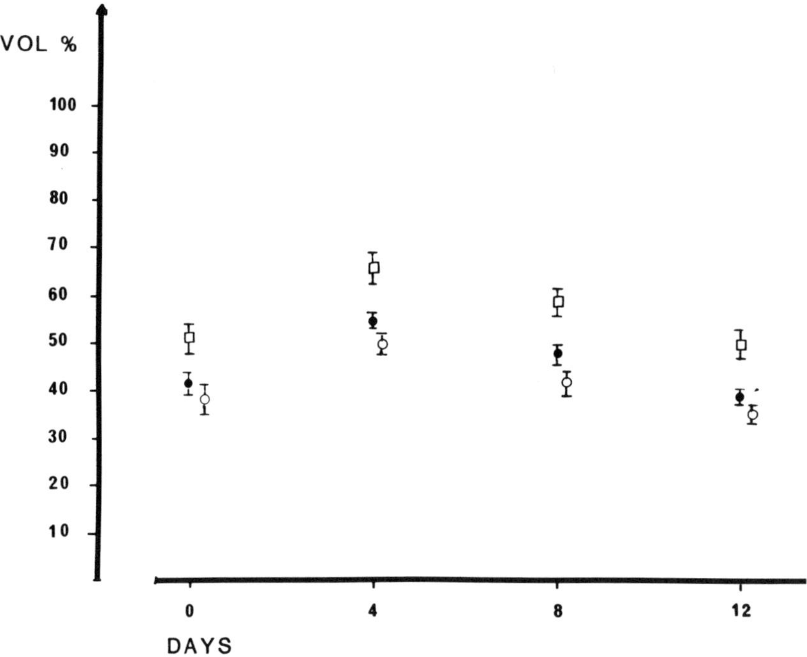

FIGURE 6. Relationship between true volume proportion of the paracortex (●), midzonal figures (□), and data from star sections (○), expressed as mean ± standard error ($\bar{x}$ ± SE), for mouse lymph nodes. The different methods produce the same pattern of variation with a nearly constant difference at each time point. The lymph nodes were stimulated repeatedly with oxazolone; 10 animals were examined per day.

μm apart in both directions gave equally good results (A. O. Myking, unpublished results). This implies that the midzonal position may be found macroscopically or with the aid of a dissecting microscope. The lymph node may be cut with a suitable knife or razor blade and the first section from either of the cut surfaces may be used for further analysis.

The purpose of the investigation was twofold: to see if the paracortical area fraction of midzonal sections could give a reasonable estimate of the paracortical volume fraction for the lymph node as a whole in (1) individual lymph nodes and (2) groups of more than six lymph nodes. To ensure an extensive variation in size and shape, nodes from two species (mice and guinea pigs) were used. They were partly stimulated with oxazolone, partly unstimulated. A total of 70 lymph nodes were examined, each representing one animal.The value of midzonal data in predicting *individual* paracortical volume fractions was, as expected, poor. In mice one-fourth of the material differed from the true values by less than 10%. The remaining three-fourths showed an overestimate of 10 to 60%. In guinea pigs only one-third of the material showed a difference of less than 10%. One-third gave underestimates of 10 to 50% and the remaining one-third overestimates of 10 to 110%. From these figures it is clear that great care should be exercised in the interpretation of paracortical size from single sections, practically the only procedure used for the study of lymph node morphology in routine histopathology.

When *groups* of lymph nodes were analyzed, midzonal data for mice showed an overestimate with a constant difference from the true value (Fig. 6). For guinea pigs there was no difference between the midzonal data and the true values (Fig. 7). For both species a higher scatter was sometimes seen in the midzonal data than in the true values.

The *practical conclusion* to be drawn from these results is that in situations where only the *variation* in paracortical volume fractions in *groups* of lymph nodes is to be studied, say in the course of antigen stimulation, it may be obtained simply by measuring the largest section of the node. The selection of the midzonal section does not require serial sectioning since macroscopic transection of the node is sufficiently accurate. The timesaving advantage of the procedure should be obvious.

Systematic Step Sections

Efforts to find a method for estimating volume fractions of lymph node compartments based on single sections have not been satisfactory because of their rather restricted applicability. The only procedure fulfilling the requirements for generality is serial sectioning or its modification, *systematic step sectioning*. With such a procedure, individual volume fractions as well as volume fractions in groups of lymph nodes may be calculated with a sufficient degree of accuracy. From a practical point of view it is important to know *how many* sections are required. In a previous investigation two related

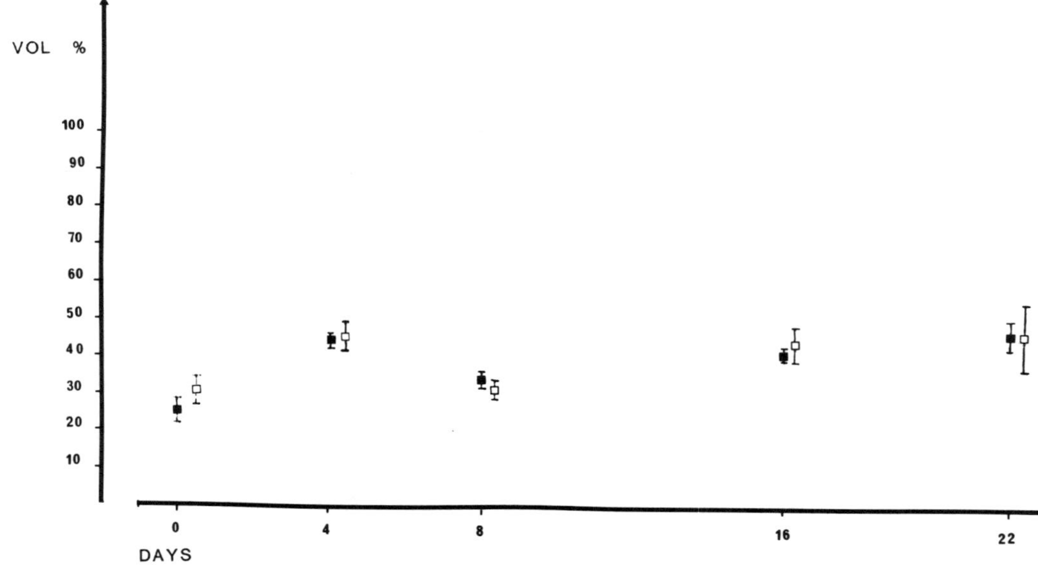

FIGURE 7. Relationship between true volume proportion of the paracortex (■) and midzonal data (□), expressed as $\bar{x} \pm$ SE, for guinea pig lymph nodes. The two methods produce the same pattern of variation, with no significant difference between the group means on the same day. On day 16 the variances are significantly different ($p < 0.01$). The lymph nodes were stimulated repeatedly with oxazolone; six animals were examined per day. (From Ref. 8, p. 377, by kind permission of the editor.)

problems were investigated[7]: (1) the minimum number of systematic sections needed to estimate an individual paracortical volume ratio within a predefined error and probability and (2) the optimum number of lymph nodes and optimum number of systematic sections from each node needed to estimate the mean paracortical volume ratio of a population with a predefined error and probability. To obtain results with general applicability, data for 70 lymph nodes (a total of more than 1100 sections representing mice and guinea pigs) were pooled and fed into a computer. As in previous studies, the basis for calculation of "true volume" proportions was systematic step sections taken at intervals of 120 μm.

To solve the *first problem,* repeated "sectioning" with a given number of sections (by varying the distance from the surface to the first section) was simulated in the computer. The error (coefficient of error) was defined as

$$\frac{\sqrt{(1/K)\Sigma_{i=1}^{K}(v_i - v)^2}}{v}$$

where K is the number of repeated trials, v_i the observed paracortical volume ratio at a defined trial, and v the true paracortical volume ratio.

With repeated sectioning (i.e., for large K) the coefficient of error stabilized at an approximately constant value. With a number of sections ≥ 2, the coefficient of error and the coefficient of variation were practically the same, i.e., the bias was negligible. For a specimen chosen randomly from a population, the following formula describes the relationship between the number of sections and the expected coefficient of error, with a probability of 0.95 fixed:

$$m = (100/\gamma_0)^{0.65} \quad (m \geq 2)$$

where m is the number of systematic sections and γ_0 is the expected coefficient of error as a percentage.

Suppose one wants to estimate the paracortical volume fraction in a lymph node with a coefficient of error less than 5% and a probability of 0.95. From the formula, seven sections are needed. With the same probability and a coefficient of error of 2.5%, 11 sections are required. Note that these figures are independent of the size of the node. *This means that with $m = 7$, for example, we can expect the coefficient of error to be 5% or less in 95% of the lymph nodes analyzed.* A solution is also given to the problem of finding m in a randomly chosen specimen with a relative error $\epsilon = (v - v)/v$ and a predefined probability (for details see Ref. 7, p. 151).

In *practical work,* the whole lymph node must be cut serially. The first section to be analyzed is taken randomly at a distance from the surface that is less than or equal to the distance between any two consecutive sections. The remaining $m - 1$ sections are taken systematically at equal distances. The area of the paracortex and of the total lymph node is measured in each of the sections taken for analysis; alternatively, point counting may be used more efficiently when many lymph nodes are to be analyzed (see next chapter).

The *second problem* concerns sample size in terms of numbers of sections and nodes needed to estimate the mean paracortical volume fraction of a population with a given relative error and probability. The problem may be to estimate the most useful size of a subsample from a larger number of nodes. The key to the solution is knowledge of the coefficient of variation $\gamma(v)$ of the paracortical volume fractions between specimens from the population. To estimate $\gamma(v)$, the paracortical volume fraction is estimated from a small random sample by taking four or five systematic sections from each node. [Fur further details concerning the calculation of $\gamma(v)$, the reader is referred to Ref. 7, p. 152.]

In *practical work* when $\gamma(v)$ is known, the most convenient combination of node number and systematic sections per node is read directly from the chart in Fig. 8. The chart is worked out for a probability of 0.95 and a relative error of 10%. Two specimen-dependent constants are involved which, from a theoretical point of view, must be calculated for each population studied. However, because of the great heterogeneity of the material used, it is not necessary to make separate calculations of these constants for other populations of lymph nodes. When a relative error (ϵ) different from 10% is desired, the corresponding number of nodes may be found by multiplying the value from the chart by $(10/\epsilon)^2$.

POINT COUNTING OR ELECTRONIC PLANAR ANALYSIS

The simple classical procedure of point counting to measure volume fractions has become less popular since the introduction of electronic equipment characterized by high precision of individual measurements of one or two dimensional structures. The resolution of these instruments is usually on the order of 0.1 to 0.001 mm^2, which is virtually infinity in comparison to most profile areas measured.[12] However, the use of electronic planar analysis in stereological work has been challenged, mainly on the grounds that it is a time-consuming procedure with an accuracy that is very often not needed.[12,13] With point counting, more efficient sampling of data is possible, allowing many specimens to be measured and thereby compensating for such important factors as the biological variation between specimens. The quality of the two methods has been compared, and results representing 40 mouse and 30 guinea pig lymph nodes stimulated chronically with oxazolone are presented

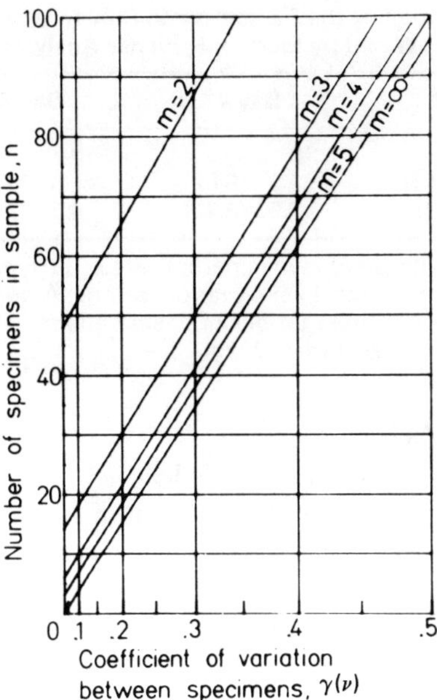

FIGURE 8. Chart for finding the minimum number (*n*) of lymph nodes to be sampled in order to estimate the mean paracortical volume fraction of a population having a coefficient of variation γ(*v*), with a relative error less than 10% and a probability of 0.95. (Reproduced from Ref. 7, p. 154, by kind permission of the editor.)

in Tables 1 and 2. These results show no differences between the group means. The variances are significantly different at some time points but show no preference for electronic planar analysis or point counting. Qualitatively (even quantitatively), the results are virtually equivalent.

The time needed for the measuring procedures was tested on 5 small nodes (day 0) and 10 larger nodes (day 12) from mice. For the smaller nodes the planimetry/point counting time ratio was averaged 4.0; for the larger nodes it averaged 2.9. Both differences were statistically significant at the 0.001 level. Model experiments indicated that both reproducibility and time consumption are clearly in favor of point counting with increasing irregularity of the profiles. Stereological studies of more complex lymph node components, such as germinal centers and medullary cords, should therefore benefit from point counting to an even greater extent than the estimation of paracortical volume fraction.

TABLE 1. Volume Fraction of the Paracortex in Mouse Lymph Nodes: Comparison of Data Obtained by Electronic Planar Analysis and Point Counting[a]

Value	Day 0	Day 4	Day 8	Day 12
n	10	10	10	10
$\bar{x}$	44.2 (43.1)	55.3 (54.2)	48.9 (48.0)	40.3 (39.6)
SD (x)	5.76 (8.9)	3.50 (3.23)	4.15 (2.90)	4.34 (5.0)

[a] Point-counting results are shown in parentheses. The lymph nodes were stimulated chronically with the contact sensitizer oxazolone. A double square lattice test system was used. The average number of coarse points falling within the area of one lymph node section was 4 ($d = 2$ cm).

CONCLUSIONS AND PRACTICAL RECOMMENDATIONS

The general solution to volumetric evaluation of focal lesions appears to be the use of systematic step sections. A theoretical consideration of the problem was recently given.[14] However, as illustrated for lymph nodes, other procedures may be applied to solve special problems. These methods, as well as the *number* of systematic step sections required, are more or less specific for a particular organ.

In lymph nodes only the requirements for estimating paracortical volume fractions have been evaluated. At present, the main field of application appears to be in research projects studying cellular kinetics after lymphocyte stimulation in vivo.

Three different principles have been described. One is based on a geometric model (star specimen model), one on a single midzonal section, and one on systematic step sections after serial sectioning of the total lymph node.

The *star specimen model* has been shown to be of rather limited practical value. It may, however, be applied for the study of inguinal lymph nodes in mice, which have been shown by reconstruction studies to have a fairly reg-

TABLE 2. Volume Fraction of the Paracortex in Guinea Pig Lymph Nodes: Comparison of Data Obtained by Electronic Planar Analysis and Point Counting[a]

Value	Day 0	Day 4	Day 8	Day 16	Day 22
n	6	6	6	6	6
$\bar{x}$	25.3 (29.4)	44.2 (42.9)	34.5 (34.8)	42.5 (41.75)	48.3 (45.1)
SD (x)	8.6 (10.3)	4.8 (5.19)	5.7 (6.6)	3.1 (1.32)	9.8 (5.3)

[a] Point-counting results are shown in parentheses. The lymph nodes were stimulated chronically with the contact sensitizer oxazolone. A double square lattice test system was used. The average number of coarse points falling within the area of one lymph node section was 17 ($d = 2.4$ cm).

ular spatial composition and in most instances to be in good agreement with the star specimen model. However, serial sectioning will be required in order to find the star point section. Systematic step sectioning is therefore a more convenient and far safer procedure. In lymph nodes the star specimen model can hardly be recommended. Its possible application to other organs such as kidneys, suprarenals, or certain parts of the central nervous system remains to be shown.

Midzonal sections are insufficient for the estimation of volume fractions in individual nodes. In groups of lymph nodes they may be used to follow alterations after lymph node stimulation and to produce pilot data for the statistical evaluation of group differences. The primary advantage of midzonal sectioning is that it is not a time-consuming procedure since its fair robustness against deviations from the midzonal position opens the possibility of macroscopic sectioning of the node.

Systematic step sections require serial sectioning of the whole lymph node. However, with suitable microtomes, the procedure may be performed with very little manual aid. The expense in terms of technical assistance may thus be reduced to an acceptable minimum. Its general applicability and high degree of precision make it the procedure of choice for most purposes. Conditions for estimating errors with a given probability have been analyzed. The research worker can thus easily find the most suitable combination of sections and nodes to fit individual requirements.

Point counting in larger series of nodes appears to be far superior to electronic planar analysis for the evaluation of volume ratios.

REFERENCES

1. Cottier H, Turk J, Sobin L: A proposal for a standardized system of reporting human lymph node morphology in relation to immunological function. Bull WHO 47:375–408, 1972.
2. Cottier H, Keiser G, Odartchenko N, Hess M, Stoner RD: De novo formation and rapid growth of germinal centers during secondary antibody responses to tetanus toxoid in mice. In: Germinal Centers in Immune Responses, edited by Cottier H, Odartchenko N, Schindler R, Congdon CC, pp. 270–276. Berlin: Springer-Verlag, 1967.
3. Meyer EM, Schlake W, Grundmann E: Comparative histologic, histochemical and histomorphometric studies of T and B cell areas in peripheral lymphoid organs of normal young adult BALB/c mice. Pathol Res Pract 164:141–156, 1979.
4. Sainte-Marie G, Peng FS: Histological observation of the deep cortex of the lymph nodes of the germ-free mouse. Ann Immunol 132D:55–63, 1981.
5. Sainte-Marie G, Peng FS: Structural and cell population changes in the lymph nodes of the athymic nude mouse. Lab Invest 49:420–429, 1983.
6. Cruz-Orive LM, Myking AO: A rapid method for estimating volume ratios. J Microsc 115:127–136, 1979.
7. Cruz-Orive LM, Myking AO: Stereological estimation of volume ratios by systematic sections. J Microsc 122:143–157, 1981.
8. Myking AO: Actual volume estimation of the paracortex in lymph nodes by morphometry. Significance of centrally placed sections. Pathol Res Pract 166:372–380, 1980.

9. Myking AO: Dynamic and morphological responses of draining lymph nodes to single and repeated applications of oxazolone to guinea-pig skin. Virchows Arch B Cell Pathol 33:155–165, 1980.

10. Myking AO: Dynamic and morphological responses of draining lymph nodes to single and repeated applications of oxazolone to the skin. Virchows Arch B Cell Pathol 31:87–102, 1979.

11. Weibel ER: Stereological Methods, volume 1, pp. 26–27. London: Academic Press, 1979.

12. Gundersen HJ, Boysen M, Reith A: Comparison of semiautomatic digitizer tablet and simple point counting performance in morphometry. Virchows Arch B Cell Pathol 37:317–325, 1981.

13. Mathieu O, Cruz-Orive LM, Hoppeler H, Weibel ER: Measuring error and sampling variation in stereology: comparison of the efficiency of various methods for planar image analysis. J Microsc 121:75–88, 1981.

14. Cruz-Orive LM: Estimating volumes from systematic hyperplane sections. J Appl Probab 22:518–530, 1985.

chapter 6

morphometry of rarely occurring cells

Kristian Ree

INTRODUCTION

To make volume density and other morphometric measurements on a cell type that is observed only sporadically during ultrastructural examination of a tissue poses certain problems. High-power magnification is necessary for definite identification of the cell according to ultrastructural criteria. An exceedingly large number of randomly selected electron micrographs would thus be needed to analyze a rarely occurring cell type, and the workload would be overwhelming. However, researchers may wish to quantify rarely occurring cell types as they frequently exert a biological effect far greater than may be supposed from their occurrence.

The present chapter outlines a method whereby rarely occurring cell types may be quantified with a reasonable workload. The methodology is exemplified by reference to a study of Langerhans cells of human epidermis.[1]

SAMPLING

Ultrathin sections for analysis are mounted on 75-mesh ($275\text{-}\mu\text{m}^2$) Formvar/carbon-coated grids. With a 75-mesh grid, a few sections will by chance lie completely unobstructed by grid bars (Fig. 1). One such section is used for morphometric analysis. Depending on the tissue to be analyzed, the entire ultrathin section in some cases serves as the reference space. In addition to the ultrathin sections mounted on 75-mesh grids, one ultrathin section is picked up from the water bath, mounted on a glass slide, and stained with toluidine blue. By looking at the ultrathin section and the corresponding semithin section in the light microscope, one can check that the trimmed area is in accordance with the predetermined sampling model (Fig. 2). The appropriate sampling model will depend on the cell type and tissue to be analyzed.

The unobstructed ultrathin section obtained by using a 75-mesh grid may in some cases enable a complete view of the entire tissue to be analyzed. For instance, by using a 75-mesh grid one obtains a complete view of a glomerulus. The sampling model in such a case is considerably simplified, particularly in relation to anisotropic properties of the tissue. By studying Langerhans cells of human epidermis, sectioned perpendicular to the skin surface, a complete view of the epidermis was obtained using 75-mesh grids (Fig. 1). Langerhans cells are dendritic cells with pseudopods extending into all strata of the epidermis (Fig. 3).

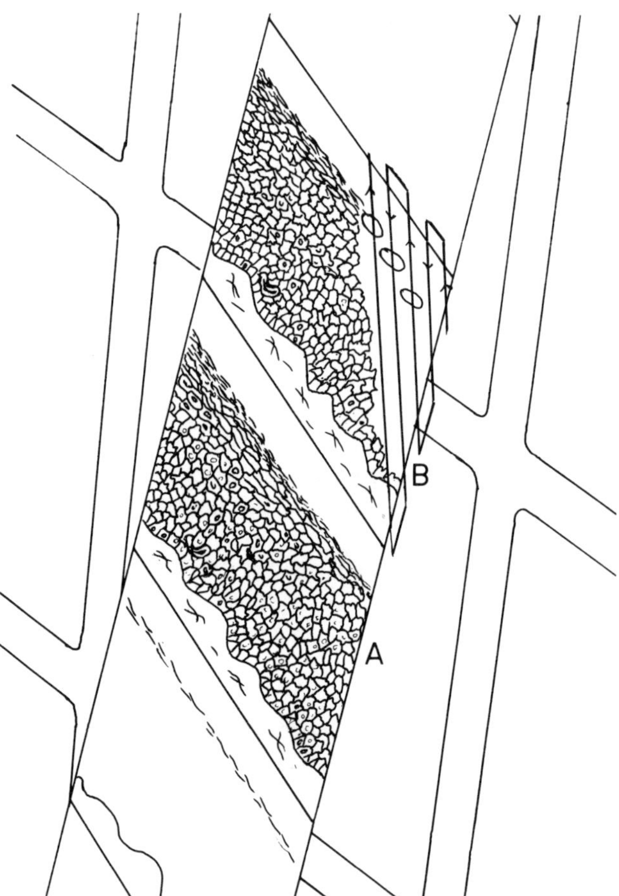

FIGURE 1. Ultrathin section of human epidermis mounted on a 75-mesh Formvar/carbon-coated grid. Section A is completely unobstructed by grid bars. In section B the parallel lines with arrowheads indicate how, during the survey of section A, the field of view is moved in a systematic manner by using the objective stage handles *one* at a time. The distance between the parallel lines must be less than the field of view.

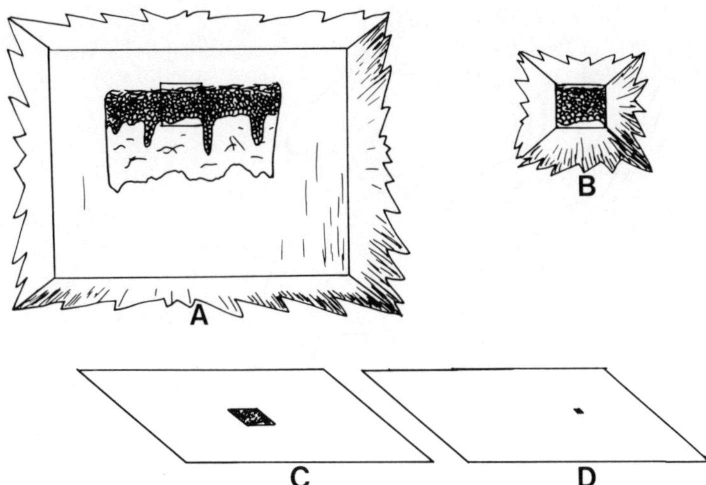

FIGURE 2. Epon-embedded human skin (A) trimmed for semithin sectioning and (B) trimmed for ultrathin sectioning. One ultrathin section is mounted on a glass slide (D) and viewed in the light microscope together with the corresponding semithin section (C).

As the complete ultrathin section occasionally is to be used as the reference space, care must be taken to avoid a systematic sampling error due to changes in the size or shape of the sections. For instance, in our study of Langerhans cells, using a section cut as a parallel trapezium would introduce a systematic error, whereas a quadrangular section would not (Figs. 1 and 3). As a rule, it is advisable to cut the sections as squares of a standardized size.

General principles of sampling in morphometry have been outlined by Weibel[2] and Gundersen and Østerby[3] and dealt with in Chapter 2 and will not be repeated here. Suffice it to say that it is desirable to use as many test animals as possible, as the biological variation between them is often greater than the variation between blocks and sections.[3] This principle also applies to infrequently occurring cells. In the latter case, however, a significant biological variation can also be assumed to be present between blocks and sections. The practical implications of this expected biological variation between blocks and sections are twofold: (1) as many test animals as possible should be included in the study, and (2) one is justified in sampling and analyzing several blocks and sections from each test animal.

TARGET CELL

Difficulties in discriminating the target cell from other cells of the tissue may put limitations on studies of this kind. Several cell types are defined ultrastructurally by the observation of specific granules in their cytoplasm. As one cannot assume, however, that the specific granules are present in every sec-

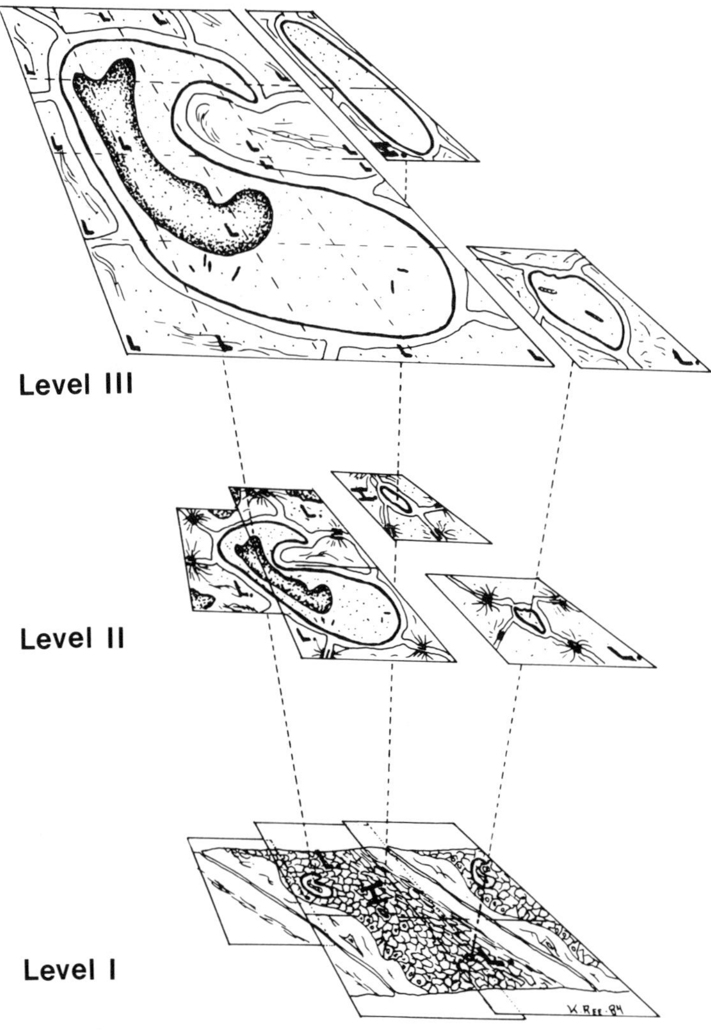

Level III

Level II

Level I

FIGURE 3. (Level I) Five electron micrographs put together giving a complete picture of the entire ultrathin section. (Level II) Target cell photographed at medium magnification. At this magnification the position of the target cell on the level I electron micrograph can be found and the cell circumscribed with a fine-point felt pen. (Level III) Sixteen electron micrographs put together to give a complete picture of the main body of the target cell. A few Langerhans cell granules are seen in two of the micrographs. The cell was thus defined as a Langerhans cell, and all the level III, II, and I electron micrographs were marked with an "L". Two cytoplasmic elements are also seen. One element contained Langerhans cell granules (marked with an L). The other element, which contained no specific features, was defined as an indeterminate cell and an "I" was put beside this element on the level III, II, and I electron micrographs.

tion that can be cut from the target cell, one cannot define the target cell by the observation of specific granules. A feature that is constantly present is needed. The cell membrane is one such feature. During the past few years it has become possible to define cells by the observation of monoclonal antibodies on the cell membrane, visualized by immunoelectron microscopy. The techniques of immunoelectron microscopy may be combined with the quantitative technique described in this chapter.

In our study of Langerhans cells[1] the problem of discriminating the target cell from the other cells of the tissue was solved by an *exclusion technique*. Keratinocytes and melanocytes can be identified in every section that can be cut from these cells. By exclusion we thus sampled nonkeratinocytes and nonmelanocytes, i.e., cells containing no desmosomes, no tonofilaments, no tonofilamentous bundles, no microfilamentous bundles, no premelanosomes, and no melanosomes. The cell population thus sampled was later subdivided into morphological entities according to ultrastructural criteria[1] (see below).

ELECTRON MICROSCOPY

The unobstructed ultrathin section chosen at random is initially photographed part by part at a primary magnification of $\times 833$. The negatives are photographically enlarged to $\times 2500$ (level I) (Fig. 3). The same section is then systematically surveyed at a primary magnification of $\times 5000$ (and $\times 10$ optical) (see legend of Fig. 1). Whenever a target cell or cytoplasmic element thereof is observed, the magnification is further increased to $\times 20,000$ and the entire target cell profile photographed part by part at a primary magnification of $\times 20,000$. The negatives are photographically enlarged to $\times 60,000$ (level III) (Fig. 3). The magnification is thereafter reduced and the same cell profile photographed at $\times 5000$ electron optical magnification. The negatives are photographically enlarged to $\times 15,000$ (level II) (Fig. 3). A complete picture of the target cell must be obtained at all levels of magnification, requiring electron micrographs to be taken with some overlap (Fig. 4).

MORPHOMETRY

With all electron micrographs available, the target cell is defined according to ultrastructural criteria and is outlined with a fine-point felt pen on the level III electron micrographs. The same target cell is then encircled on the level II electron micrographs. With the aid of the level II electron micrographs, the position of the target cell on the level I electron micrographs is found and the cell is circumscribed with a fine-point felt pen (Fig. 3).

To ensure that no target cell element was left out during the systematic survey of the section at $\times 5000$, the same section should be surveyed a second time. The reexamination of the section should be performed with all electron micrographs pertaining to that section available next to the operator of the electron microscope. Since the target cell has been circumscribed on

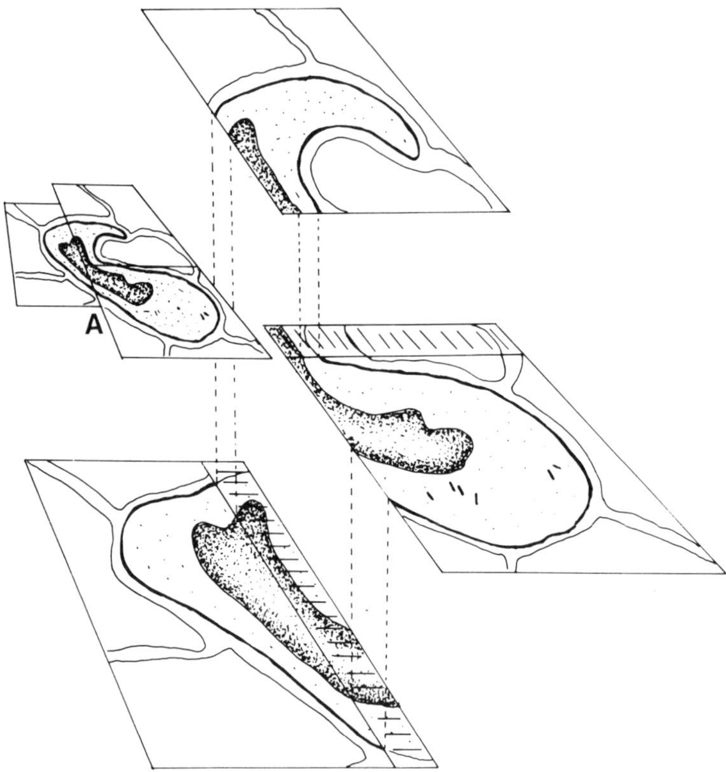

FIGURE 4. Three electron micrographs fitted together to give a complete picture of the target cell (A). On each electron micrograph a line is drawn to point out the overlapping area.

the electron micrographs prior to the reexamination, any cell element left out during the first screening is quickly recognized.

In our study of Langerhans cells, all level III electron micrographs were scrutinized for Langerhans cell granules. Whenever one or more granules were observed, the cell was defined as a Langerhans cell and the cell element marked with an "L" on the level III, II, and I electron micrographs (see legend of Fig. 3).

The electron micrographs are fitted together and a line drawn to show which area on each electron micrograph is to be excluded from the morphometric analysis (Fig. 4). The actual point counting can now be performed. The appropriate lattice d spacing of the double lattice test grid must be decided on after some pilot measurements. For the Langerhans cells we found a lattice d spacing of 0.5 cm, with 36 fine points per coarse point, suitable for both the level I and level III measurements (Figs. 5 and 6).

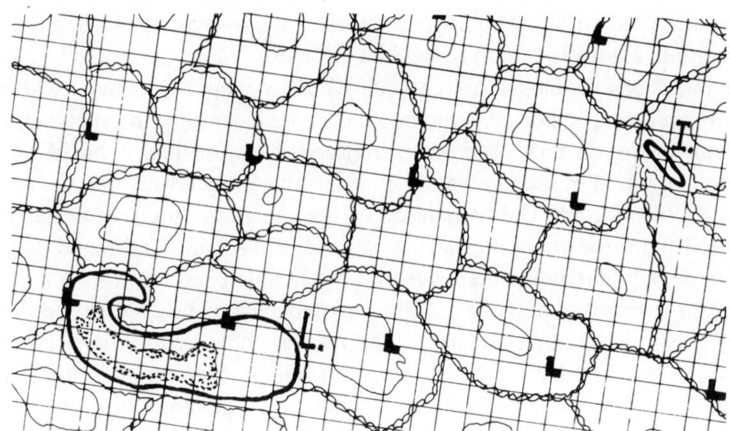

FIGURE 5. Drawing of a level I electron micrograph with a Langerhans cell (L) and an indeterminte cell element (I) encircled with a fine-point felt pen. A double lattice test grid is placed on the electron micrograph.

DISCUSSION

The use of 75-mesh Formvar/carbon-coated grids is a prerequisite for applying the methodology described in this chapter. One problem encountered when using these grids is that during electron microscopic examination of the tissue the image started to drift. Drifting is usually caused by lack of stability in the Formvar film supporting the section. This may be counteracted by using a low beam current and exposing the Formvar-coated grids with substantial amounts of carbon (visible on white paper). Both the Formvar and the carbon will reduce the resolution of the specimen. For most magnifications

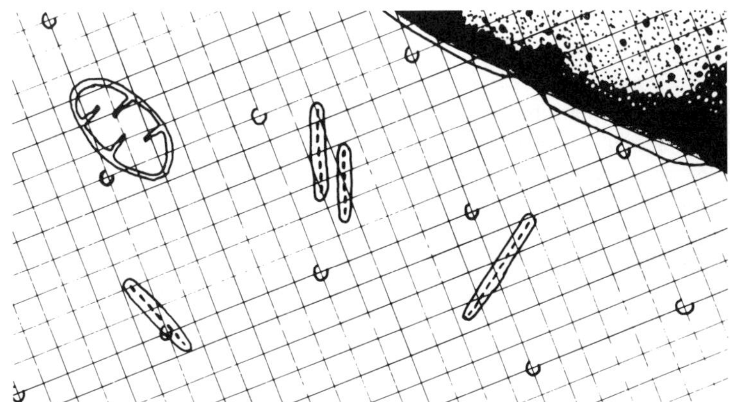

FIGURE 6. Drawing of a level III electron micrograph depicting four Langerhans cell granules in the cell cytoplasm. A double lattice test grid is superimposed.

generally used to study biological processes, however, the resolution obtained with Formvar/carbon more than suffices.

Pfeifer[4] has developed a technique whereby autophagosomes, which are rarely occurring spherical structures, may be quantified morphometrically without having to take electron micrographs. The method is based on analyzing the ultrathin section that by chance completely covers one square hole of a 100-mesh grid. The boundaries of the square are used as the reference space. Pfeifer measured the sizes of several grid squares and found that they differed. Individual measurements of the grid hole parameters were therefore necessary. For this purpose a digital counter was connected to the driving mechanism of the objective stage. The fluorescent screen of the electron microscope was fitted with regularly spaced points, thus enabling direct measurements of the spherical target structure (autophagosomes). With some minor adaptations, Pfeifer's technique may be used to study rarely occurring cell types as well as other rarely occurring organelles. For solving specific problems, the technique is superior to the method described in the present chapter inasmuch as it is more time- and cost-effective. Time/cost effectiveness is an important factor when planning a morphometric study. The present technique is probably more versatile, enables a complete morphometric profile to be obtained for the cell and all its constituents, and does not require any prearrangements on the electron microscope.

Morphometric calculations are based on random sampling of data. A negligible error is introduced into such calculations if by chance a structure is counted twice due to incidental overlapping of two or more micrographs. The methodology described here requires, however, that every micrograph be photographed with some overlap. Counting the same structures twice would thus become a regular phenomenon, and the error introduced cannot a priori be assumed to be negligible. By fitting the electron micrographs together and drawing a line to show which area of the micrograph is to be analyzed, any systematic sampling error at this stage is minimized (Fig. 4). Correct sampling at all stages is crucial in morphometry in general and particularly so when doing calculations on rarely occurring structures. The combined examination of semithin sections and ultrathin sections, as well as the mounting of sections on 75-mesh grids, allows a fully controllable sampling technique to be applied.

REFERENCES

1. Ree K: Reduction of Langerhans cells in human epideris during PUVA therapy. A morphometric study. J Invest Dermatol 78:488–492, 1982.
2. Weibel ER: Stereological Methods, volume 1, Practical Methods for Biological Morphometry. London: Academic Press, 1979.
3. Gundersen HJG, Østerby R: Optimizing sampling efficiency of stereological studies in biology—or "Do more less well!" J Microsc 121:65–73, 1982.
4. Pfeifer U: The evaluation of large test fields for morphometric studies in electron microscopy. Pathol Res Pract 166:188–202, 1980.

part 3

surface measurements

This part comprises three chapters dealing with component surface densities in a reference volume (S_V) and on a reference surface (S_S).

Chapter 7, by Halgunset, presents an ingenious approach to an important problem: how to estimate the average volume of irregularly shaped particles. In this case the particles are mononuclear phagocytes in cell culture, and Halgunset proposes estimating average volumes from their relative and absolute areas of attachment to the substratum.

Mayhew, Middleton, and Ross (Chapter 8) take advantage of recent developments in intestinal morphometry to derive efficient estimates of surface amplification factors due to villi and microvilli. The methods are based on surface densities on a surface (S_S) obtained from transverse sections only. Lack of random and independent orientation is accounted for by use of modified stereological formulas, and absolute surface areas are used to compare normal and fasted rats.

Mayhew and Reith (Chapter 9) treat the problem of membrane image loss due to oblique sectioning, emphasizing that this is but one of several technical biases influencing estimates of membrane surface densities. They recommend calculating correction factors to suit membrane species and individual experimental needs rather than abstracting what may be inappropriate correction factors from the literature. The methods are illustrated using several membranes, including those of nuclei and mitochondria.

chapter 7

volume estimation of irregular structures: human monocytes in culture

Jostein Halgunset

INTRODUCTION

In biology, the basic structural and functional unit is the cell, and so the average cell will often be the natural reference unit for morphometric data. Increased informational value can be obtained by relating the structural quantities to the mean cell volume. An alteration of the nuclear volume may be a morphological correlate of fundamental cellular processes because the size of the nucleus depends on its DNA content. However, a biologically important increase of the nuclear volume may fail to produce a detectable increase of the nuclear volume density if there is concomitant hypertrophy of the cells. In this situation, the average nuclear volume per cell would be a less ambiguous parameter. Absolute measures, the amount of a structure per average cell, cannot be obtained directly by stereological methods, but component density values can be converted into absolute values by taking into account the average cell volume. The average cell volume itself may also be a biologically meaningful parameter, as changes of the cell volume accompany several physiological and pathological processes.

There is no simple, generally valid method for the estimation of average cell volume. Sometimes, one may even have to use laborious reconstruction procedures based on serial sectioning. In other instances, some additional information is available or a few reasonable assumptions or approximations can be made that permit the problem to be solved in an easier way. Isolated cells in suspension, like blood leukocytes or cells in suspension culture, tend to assume nearly spherical shapes. Their volumes can be calculated to an acceptable degree of accuracy by applying the formula for the volume of a sphere or an ellipsoid of revolution. In many cases the cells are irregular but their nuclei can be considered as spheres. Then the average nuclear volume

$\bar{v}(n)$ can be estimated from the profile diameter distribution. Provided each cell contains a single nucleus, the average cell volume is estimated by combining $\bar{v}(n)$ and the nuclear volume density $V_V(n, c)$:

$$\bar{v}(c) = \frac{\bar{v}(n)}{V_V(n, c)} \tag{1}$$

VOLUME ESTIMATION FOR CELLS IN MONOLAYER CULTURE

The simplest way to obtain an estimate of the volumes of cultured cells is to bring them into suspension by treatment with trypsin or EDTA, followed by diameter measurement. It is generally believed that such procedures result in a simple change of the cells' shape without appreciable alteration of their volume. However, for morphometric analysis of cells in monolayer culture, it may be desirable to preserve their original shape by in situ fixation. Thus, a disadvantage of the trypsinization method is that the volume estimation cannot be performed with the same cells used for stereological morphometry, but must be done in parallel cultures. Furthermore, certain cell types are virtually impossible to loosen from the substratum. In particular, this is the case for human monocyte-derived macrophages.

Ko and Koestner[1] described a formula for calculating the volumes of individual rat glioma cells in culture by measuring cell diameters and heights in the light microscope, approximating the cell shape to a geometrically simple model. Apart from being laborious, a method like this is probably not very accurate or precise, as most cell types display great variation in shape.

When cell monolayers are observed by light microscopy using phase contrast optics, the dimensions of the nuclei are directly measurable. If the nuclei are spherical and the cells are mononuclear, the method based on Eq. (1) may be applied. However, it appears that such nuclei tend to be somewhat flattened. The horizontal diameters are readily seen, but the vertical diameter is inaccessible to measurement by direct microscopy. Thus, the nuclear volume is not easily calculated. Moreover, some cells have irregular nuclei, and in many cases a fraction of the cells have more than one nucleus.

The nuclear volume is a three-dimensional parameter, which cannot be determined by measuring the area of its two-dimensional projection. However, there exists a well-defined cellular feature that is two-dimensional and perpendicular to the axis of projection: the cell attachment area, i.e., the area of contact between the cell and the substratum. Insofar as this area corresponds to the observed vertical projection of the whole cell onto the substratum, it is measurable in the light microscope to any desired degree of precision (limited only by the resolution of the microscope). In this area the cytoplasmic membrane is molded on the flat substratum and is therefore essentially plane. When the cells are fixed in situ, this shape will be retained during subsequent processing for transmission electron microscopy (TEM). Whenever the sectioning plane cuts the cell membrane of the attachment area, the resulting trace will appear as a more or less straight line. It is therefore readily recognized in the electron micrographs. By conventional stereological procedures the surface density of the "attachment membrane" relative to the cell volume may be estimated. In analogy with Eq. (1), the following relation exists between the average cell volume $\bar{v}(c)$, the average attachment area per cell $\bar{s}(a)$, and the attachment membrane surface density $S_V(a, c)$:

$$\bar{v}(c) = \frac{\bar{s}(a)}{S_V(a, c)} \tag{2}$$

$S_V(a, c)$ is the surface density of the idealized plane lower surface of the cells, corresponding to the area of the vertical projection of the cells.[2]

PRACTICAL PROCEDURE

The average attachment area per cell is readily determined by planimetry of the light microscopic image of the monolayer. Because of overlapping of cells the exact tracing of the outlines of each individual cell may often be difficult. But as the relevant parameter is the average attachment area in the cell population, it suffices to determine the total attachment area for a sample of cells and the number of cells that contribute to this total. This can be done very conveniently by point counting. A test area A_T containing an array of P_T test points is applied to the image of the culture, ensuring independence in both position and orientation (Fig. 1). The number of points $P(c)$ that fall on the cells is recorded. The cell number $N(c)$ is obtained using an unbiased counting rule by counting all cells entirely within the test area as well as those that touch the broken line, but excluding all cells touching the unbroken line.[3] Then $\bar{s}(a)$ is estimated from

$$\bar{s}(a) = \frac{\sum\limits_{i=1}^{n} P(c)_i}{\sum\limits_{i=1}^{n} N(c)_i} \frac{A_T}{P_T M^2} \tag{3}$$

where M denotes the final magnification and n is the number of fields examined. The test fields must be randomly sampled over the entire culture, e.g., by a systematic random procedure. If the cells are grown on the bottom of a culture dish, they should be prevented from attaching at the most peripheral part of this area, which is difficult to evaluate because of reflection and scattering of the light by the side walls. As an alternative, the cells may be grown on coverslips that can be taken out of the dish for microscopy.

After light microscopy examination, the cell monolayer is further processed for TEM. During embedding and sectioning the cells must be random-

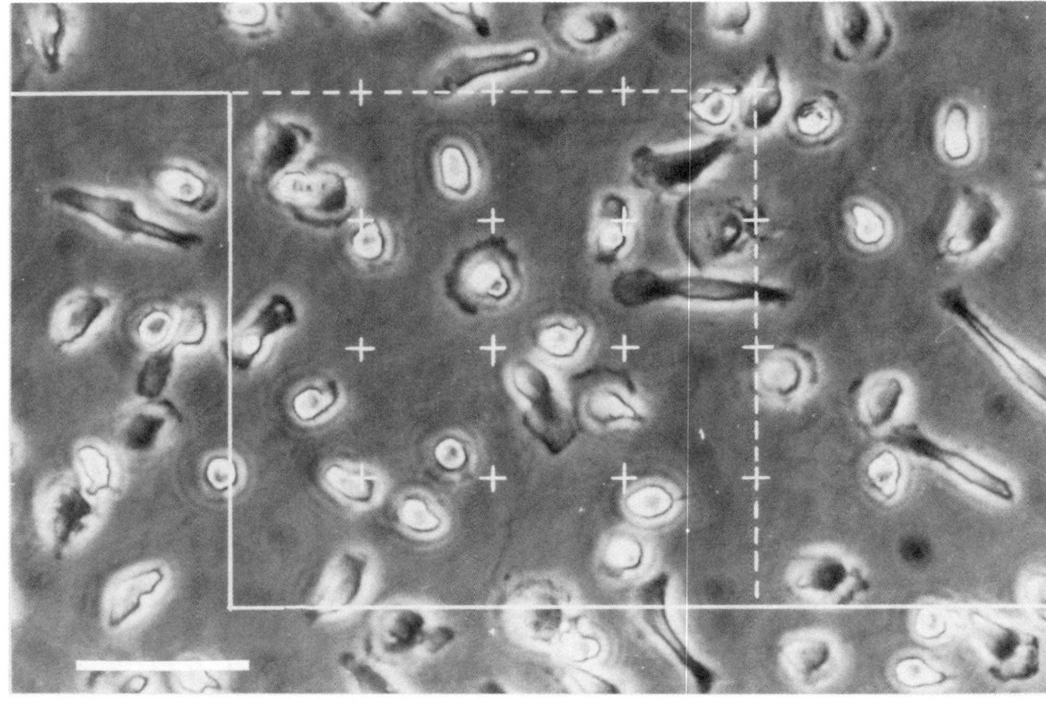

FIGURE 1. Phase contrast light micrograph of human monocytes fixed in situ after 2 days of culture, overlaid with the grid used for estimation of average attachment area. Bar, 50 μm.

ized with respect to localization and orientation. This can be achieved to a large extent by loosening the fixed cells from the substratum, using propylene oxide to dissolve the polystyrene, followed by centrifugation. There is some controversy as to whether this procedure ensures proper randomization. It has been claimed that differential sedimentation and preferential orientation may take place.[4] However, the fixed monolayer generally comes off the dish as cellular sheets of varying sizes rather than single cells. These sheets intermingle, thereby restricting the movement of individual cells during centrifugation. Thus it seems unlikely that important systematic differences occur. There is a tendency for neighboring cells to lie in parallel, but more widely separated cells do not display such interdependence of orientation.

If the electron micrographs can be considered a random sample, the attachment area surface density may be estimated by conventional stereological methods. A point-and-line grid with P'_T test points and a total test line length of L_T is applied to the micrographs (Fig. 2). The number of intersections $I(a)$ between test lines and attachment membrane trace is recorded. Identification of the attachment membrane is facilitated by the presence of a continuous electron-dense line representing proteinaceous material adsorbed onto the substratum during culture (see Fig. 2, arrowheads). However, the relevant feature is the idealized plane lower surface of the cells, corresponding to the projection of the cells onto the substratum. This must be taken into account when performing the intersection count; the membrane trace should be "smoothed," ignoring smaller irregularities. The number of test points $P'(c)$ that fall on cellular profiles is further recorded. $S_V(a, c)$ is estimated from

$$S_V(a, c) = \frac{\sum_{i=1}^{n} I(a)_i}{\sum_{i=1}^{n} P'(c)_i} \frac{2P'_T M'}{L_T} \tag{4}$$

where M' denotes the final magnification and n is the number of micrographs examined. Finally, $\bar{v}(c)$ is estimated from Eq. (2).

With this method there is no need for any assumptions concerning cellular or nuclear shape, except that the contours seen by light microscopy are supposed to outline the cells' area of attachment to the substratum. This means that no part of a cell should project over an area of bare plastic. A large number of electron micrographs have shown that this is very nearly true for many cell types.

The performance of this method was tested using an established cell line, NHIK 3025. Four parallel cultures were fixed in situ and the average cell volume estimated as described above. Two additional culture wells were trypsinized, followed by measurement of the cell diameters. The results obtained by the two methods are in relatively close agreement, as shown in Fig. 3.

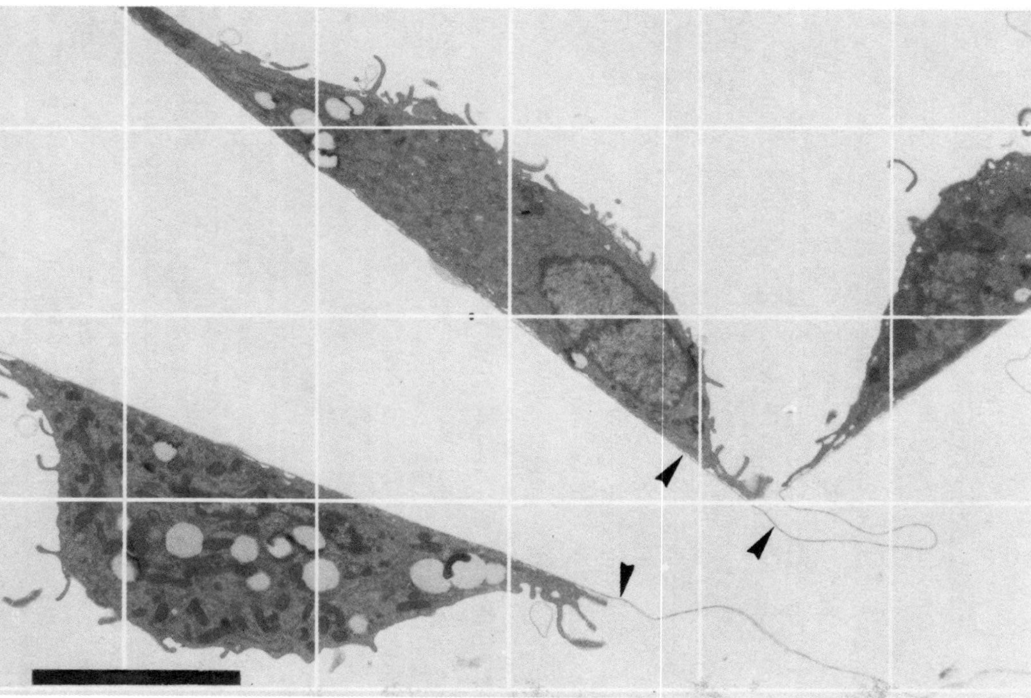

FIGURE 2. TEM of the monocyte culture shown in Fig. 1, overlaid with the grid used for estimation of attachment area surface density. Arrowheads point at the electron-dense line, which serves as a marker of the attachment membrane. Bar, 5 μm.

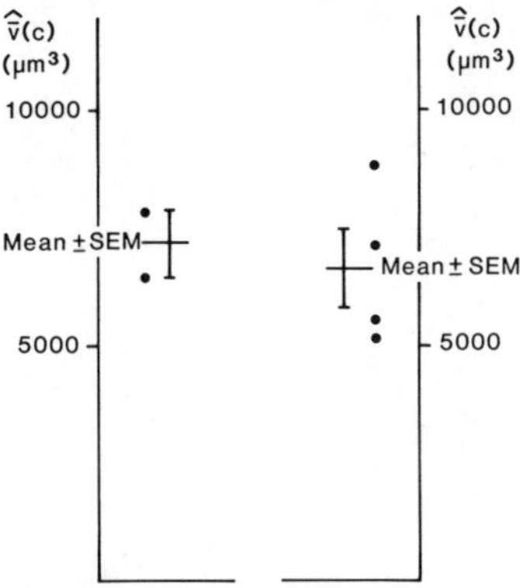

FIGURE 3. Estimation of average cell volume (NHIK 3025 cells) by trypsinization and diameter measurement (left) and by the stereological method described in this chapter (right). (Reproduced from Ref. 2 with permission of the publisher.)

VOLUME CHANGES IN HUMAN PERIPHERAL BLOOD MONOCYTES DURING CULTURE

When kept in monolayer culture, peripheral blood monocytes develop into macrophagelike cells, undergoing a series of morphological and functional changes.[5] Stereological methods are well suited for characterizing this kind of process. Since one of the most conspicuous changes that takes place is a dramatic increase in cell size, it would be desirable to express the morphometric data in absolute values. Macrophages have irregular shapes and irregular or multiple nuclei and are virtually impossible to loosen from the substratum. Thus, none of the traditional methods for volume estimation would be of any use. Kristensen and Papadimitriou[6] have devised a method based on scanning electron microscope (SEM) photogrammetry for estimating the volume of individual cells in monolayers. This would be a very laborious way to obtain average values.

An estimate of average monocyte-macrophage cell volume is readily obtained by the stereological method outlined above. An experiment was performed in which six parallel monocyte cultures were fixed at each of seven time points: 90 min and 1, 2, 4, 6, 8, and 10 days after initiation of the culture. Ten random phase-contrast light micrographs were taken of each culture after fixation. The stepwise dehydration was followed by treatment with propylene oxide, which made the monolayer loosen from the dish. The cells were then centrifuged into the tip of BEEM capsules. One section was

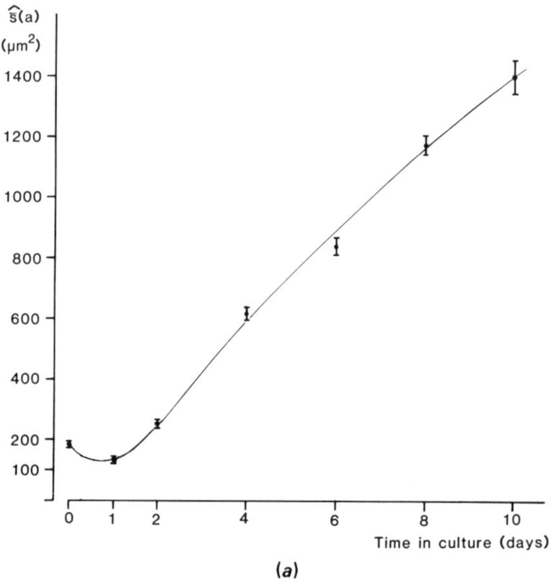

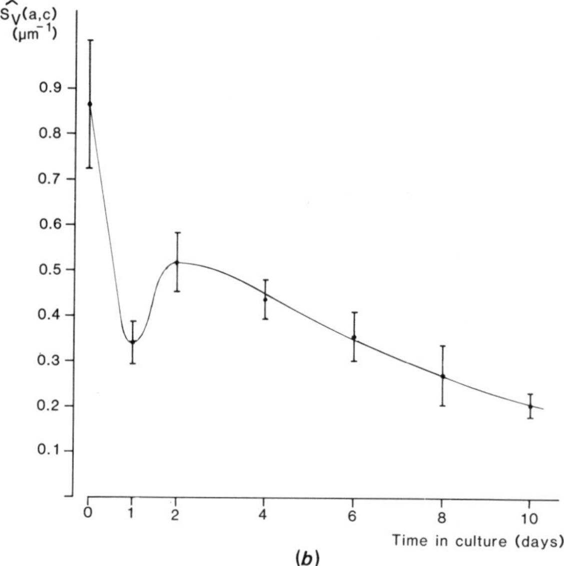

FIGURE 4. Results of the stereological method for volume estimation applied to cultures of human blood monocytes. (*a*) Average attachment area per cell; (*b*) attachment membrane surface density.

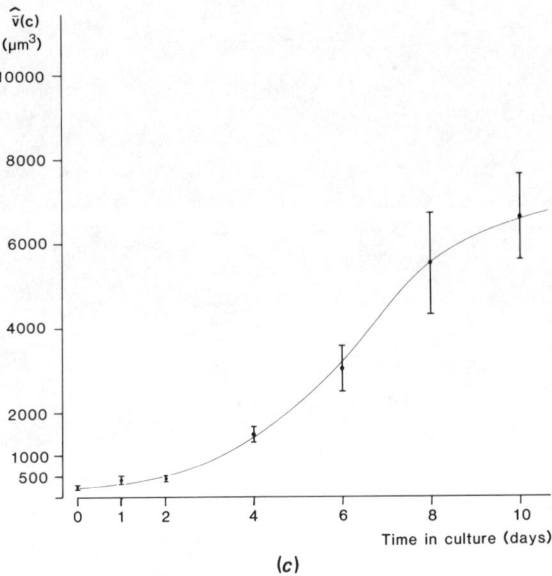

FIGURE 4. Results of the stereological method for volume estimation applied to cultures of human blood monocytes (*Continued*). (c) Average cell volume. The means ± SEM (*n* = 6) are shown as a function of the duration of the culture period. [Part (c) reproduced from Ref. 2 with permission of the publisher.]

examined from each culture, 10 systematic random electron micrographs being taken from each section. The two sets of micrographs were overlaid with the appropriate test grids, and the counting was performed. The resulting data are shown in Fig. 4.

On day 0, i.e., after 90 min of adherence, the average attachment area was 187 μm^2 (SEM = 5.7 μm^2) and the attachment membrane surface density was 0.8664 μm^{-1} (SEM = 0.1405 μm^{-1}). The final estimate of the average cell volume (the mean of the estimates in the individual wells) was 235 μm^3 (SEM = 34 μm^3), which is the volume of a sphere of diameter 7.66 μm. This value is in close agreement with the results of Schmid-Schönbein et al.,[7] who found a mean monocyte diameter of 7.47 μm. One day after initiation of the culture there was some decrease of $\bar{s}(a)$, but this was accompanied by a large reduction of $S_V(a, c)$ and the average cell volume was found to increase a little. On the second day, $\bar{s}(a)$ had increased to reach its initial value, and there was also an increase of $S_V(a, c)$. Thus, there was a small increase of the cell volume from day 1 to day 2. During the rest of the experiment, there was a continuous increase of $\bar{s}(a)$ while $S_V(a, c)$ was steadily decreasing. The resulting estimates of $\bar{v}(c)$ show an impressive 28-fold increase during the

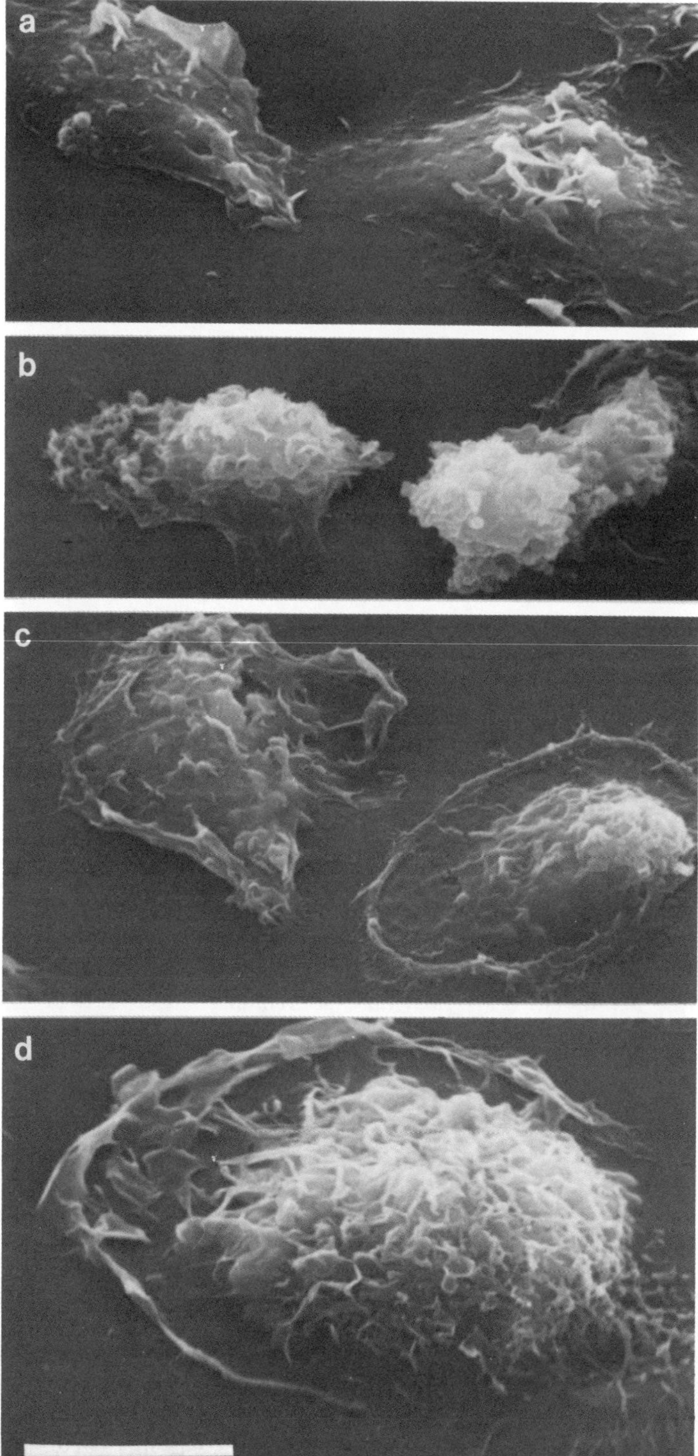

FIGURE 5. Scanning electron micrographs of cultured human blood monocytes, fixed in situ at 90 min and 1, 2, and 4 days of culture (a, b, c, d respectively). Micrographs show typical examples of cells, illustrating morphological changes demonstrated quantitatively in Fig. 4. All cells shown at same magnification. Bar, 10 μm.

10-day period. Part of this increase is accounted for by the progressive fusion of cells into multinucleated giant cells.

The estimates of $\bar{s}(a)$ and $S_V(a, c)$ also carry information about the cell shape, the degree of cell spreading. In fact, $S_V(a, c)$ is related to the inverse of the average cell height. The three curves shown in Fig. 4 demonstrate a sequence of shape changes that take place in monocytes during culture. The initial attachment of the cells is accompanied by pronounced flattening, the average cell height being estimated as $(0.8664 \ \mu m^{-1})^{-1} = 1.15 \ \mu m$. During the following 24 h the cells retract, assuming a more ''rounded'' shape. The average height on day 1 is 2.90 μm. One day later, the cells have spread again to cover a larger area of the substratum but are somewhat less flattened than at the 90-min stage. Thereafter, there is a continuous increase of the cells' extension as well as their height. These changes are illustrated in Fig. 5. The shape changes during the first 2 days of culture probably represent the morphologic counterpart of the adaptation to in vitro conditions.

REFERENCES

1. Ko L-W, Koestner A: Morphologic and morphometric analyses of butyrate-induced alterations of rat glioma cells in vitro. J Natl Cancer Inst 65:1017–1027, 1980.
2. Halgunset J: Stereological estimation of average cell volume in monolayer culture by combined light and electron microscopy. J Microsc 135:325–336, 1984.
3. Gundersen HJG: Notes on the estimation of the numerical density of arbitrary profiles: the edge effect. J Microsc 111:219–223, 1977.
4. Mayhew TM, White FH: Ultrastructural morphometry of isolated cells: methods, models and applications. Pathol Res Pract 166:239–259, 1980.
5. Ødegaard A, Viken K-E, Lamvik J: Structural and functional properties of blood monocytes cultured in vitro. Acta Pathol Microbiol Scand Sect B 82:223–234, 1974.
6. Kristensen SEL, Papadimitriou JM: The measurement of cell volume and surface area by SEM photogrammetry. J Microsc 124:155–161, 1981.
7. Schmid-Schönbein GW, Shih YY, Chien S: Morphometry of human leukocytes. Blood 56:866–875, 1980.

chapter 8

dealing with oriented surfaces: studies on villi and microvilli of rat small intestine

T. M. Mayhew, Carolyn Middleton, and G. A. Ross

INTRODUCTION

In the past two decades there have been several advances in our understanding of the rate and control of cell production in the small-intestinal mucosa.[1-5] In order to calculate total villus transit times from cell production rates, it is useful to have some indication of villus surface area at various sites along the intestine.[2] Estimates of surface areas are valuable in other contexts. Thus, physiological gradients of glucose utilization correlate well with anatomical gradients of villus surface area.[6] Moreover, villus surface area correlates better than villus height with the absorptive capacity and enzymatic activity of the mucosa.[7] There is also evidence that villus surface area changes with experimental treatment and during disease.[6-12]

Previous studies of the intestinal surface encountered practical problems related to the fact that the mucosa has a complex three-dimensional organization. Attempts to define simple geometric models that approximate the form of the average villus[2,6,7] are unsatisfactory because villus size and shape vary with location and with age.[13,14] Estimating villus surface amplification factors by combined transverse and longitudinal sectioning[6,8,11] is time-consuming, especially when data are required for different intestinal segments from adequate numbers of animals. An alternative descriptor of mucosal morphology, villus height, has proved useful for investigating structural gradients within animals[5,8,15] but has only limited value for comparing experimental groups[15] or for correlating with functional parameters.[7]

All of these problems can be circumvented by a methodology based on recent analyses of the orientation characteristics of surfaces in the rat small

The authors are grateful to Professor E. J. Clegg for his continued support and encouragement of this work.

intestine.[16] In this animal, the primary mucosal surface is anisotropic (it has a preferred longitudinal orientation because it is essentially cylindrical) but the overall villus surface is isotropic (shows no preferred direction of orientation in space). Because of these orientation properties, valid and precise estimates of villus and microvillus surface areas can be obtained by analyzing only transverse sections.[16] Furthermore, the method is independent of variations in villus size and shape provided the above orientation conditions are met.

In this chapter we describe an experimental protocol for estimating the relative and absolute surface areas of villi and microvilli by combined light and electron microscopy. The stereological procedures are illustrated by referring to recent studies of the small-intestinal mucosa of normal (nonfasted) and fasted rats.

EXPERIMENTAL PROTOCOL

The following considerations apply strictly to the small intestines of normal and fasted rats. For these groups, there is experimental evidence that the necessary orientation conditions are satisfied.[11,16]

Definition of Relevant Surfaces

The *primary* mucosal surface may be defined at the light microscopic level as the boundary between the bases of villi and the openings of crypts.[11] This surface corresponds to that of a long folded cylinder and can be imagined as the surface that would be present if there was no enlargement by villi. The *secondary* surface is that due to villi. At the ultrastructural level, this is represented by the boundary running parallel to the luminal plasmalemma of enterocytes at the bases of microvilli.[11] It is the surface that would be present if there were no microvilli. The *tertiary* surface is that due to microvilli.

Villus and microvillus surface amplification factors are merely stereological surface densities on a surface, for which the symbol S_S is the conventional notation.[17] The degree to which villi amplify the primary mucosal surface can be expressed as $S_S(v, p)$ and the amplication by microvilli as $S_S(m, v)$. Provided the definition of the villus surface is consistent at the light and electron microscopic levels, the overall amplification due to microvilli $S_S(m, p)$ can be estimated by the product $S_S(v, p)S_S(m, v)$. Here $S_S(v, p)$ is estimated by light microscopy using semithin sections and $S_S(m, v)$ by electron microscopy using ultrathin sections.

Since amplification factors are just ratios of surfaces, "absolute" surface areas can be estimated only if one of the relevant surfaces is known. Fortunately, it is possible to achieve this by estimating the circumference and length of the primary mucosal cylinder in each animal. Letting $S(p)$ denote this absolute reference surface, we can calculate the total villus surface per animal $S(v)$ from the expression

$$S(v) = S_S(v, p)S(p) \tag{1}$$

and the total surface due to microvilli $S(m)$ from

$$S(m) = S_S(v, p)S_S(m, v)S(p) \tag{2}$$

Stereological Principles

Counting intersections I between the sectional images of the relevant surfaces and independently positioned and orientated test lines is an efficient way to estimate surface densities from tissue sections.[18,19]

For a given region of the small intestine, the best way to estimate $S_S(v, p)$ is to count test intersections on transverse sections.[16] To account for slight departures of the primary mucosal circumference from circularity, a simple quadratic test lattice may be superimposed on transverse sections and intersection counts summed over "horizontal" and "vertical" test lines. The fundamental stereological principle relating intersection counts to surface density is then[16]

$$S_S(v, p) = \frac{4}{\pi}I_I(v, p) \tag{3}$$

where $I_I(v, p)$ is the ratio of total intersections through villi, $I(v)$, to those through primary mucosa, $I(p)$. The numerical coefficient $4/\pi$ arises because of the surface orientation properties and the (transverse) direction of sectioning of the intestine.[16]

Using ultrathin transverse sections, and given that both the villus and microvillus surfaces are isotropic,[16] the surface density $S_S(m, v)$ may be estimated using

$$S_S(m, v) = I_I(m, v) \tag{4}$$

where $I_I(m, v)$ is the ratio of total intersections through microvilli, $I(m)$, to total intersections through villi, $I(v)$. For this purpose, it will be necessary to use a magnification that offers adequate resolution of microvilli[11,20] and to bear in mind other potential sources of technical bias[21,22] (also see Chapter 9, this volume).

The boundary length (circumference) of the primary mucosa, $B(p)$, can be estimated from intersections with horizontal and vertical test lines, $I(p)$, using the relationship

$$B(p) = \frac{\pi}{4}hI(p) \tag{5}$$

where h represents the test lattice spacing in terms of the specimen dimensions, i.e., taking into account the specimen magnification. This estimate of $B(p)$ can be made on the semithin transverse sections.

For a segment of intestine of length L, the "absolute" surface of the primary mucosa, $S(p)$, can then be calculated from

$$S(p) = LB(p) \qquad (6)$$

and $S(v)$ and $S(m)$ can be estimated for the same length of intestine from Eqs. (1) and (2), respectively.

Tissue Sampling

One may distinguish two sorts of sampling regime depending on whether it is required to (1) study structural gradients of surface amplification at various sites along the small intestine or (2) estimate absolute surface areas in the intestine as a whole.

In the first case, pieces of tissue may be excised at desired locations. For example, pieces might be taken from each animal at sites corresponding to 10, 20, 30, . . . , 90% of total intestinal length from some convenient starting point, say the pylorus.[2] This approach is preferable to removing samples at, say, 10-cm intervals,[6] since individual intestinal lengths may vary appreciably even in animals of similar body weight.

In the second case, each intestine can be divided into a desired number of segments of roughly equal length. From each segment, simple random or systematic random samples of tissue are removed for processing. The selection procedure should be reproduced for each segment in turn.

After processing for electron microscopy, each piece of intestine will provide a number of suitably oriented tissue blocks. From the pool of blocks representing a given segment, those to be sectioned can be selected randomly (e.g., by lottery). For light microscopy, a single complete arbitrary transverse section can be cut from each block for intersection counting[16] (see Fig. 1).

Parallel ultrathin sections for electron microscopy may be cut from the same blocks, if necessary after trimming down the block face so that a random *sector* of the intestinal cross section remains. Micrographs depicting villus profiles may then be recorded using a systematic random sampling regime[18] (see also Fig. 1). This is preferable to recording electron micrographs at the tips and bases of villi[11] if data characterizing the villus surface as a whole are required, as is the case in formula (2).

We illustrate results obtained using the above methods to study mucosal architecture in nonfasted and fasted rats.

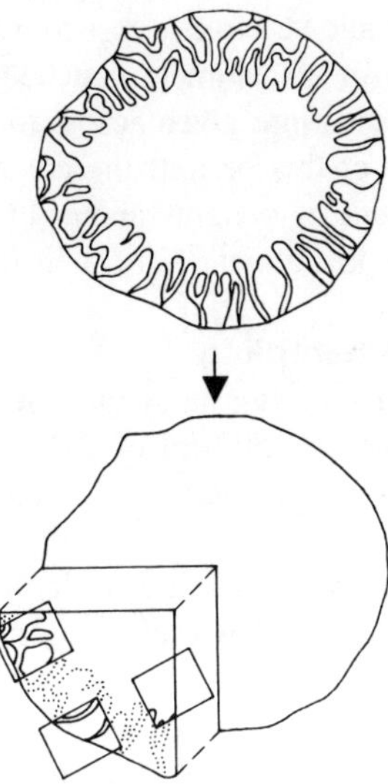

FIGURE 1. Tissue sampling for light and electron microscopy. The upper diagram illustrates a complete transverse section showing images of the primary and villus surfaces of the mucosa. For ultrathin sectioning the block is first trimmed down to a random sector (bottom diagram) from which systematic random micrographs are sampled in order to quantify microvilli on villi.

MATERIALS AND METHODS

Animals

Adult female rats of the black and white hooded Lister strain were employed. They were reared in normal laboratory conditions and killed in groups at the same time of day to minimize diurnal variations. Two experiments are described. In the first, six animals of 260–300 g body weight were fed a standard pellet diet and allowed drinking water ad libitum. These animals were used to study structural gradients of villus morphology from the pylorus to the ileocecal junction. Some information on microvillus surface areas was also obtained.

In the second experiment, we wished to investigate the effects of fasting on villus surface areas. For this purpose, six rats of 260–310 g body weight were deprived of food, but allowed free access to water, for 48–51 h prior to sacrifice. A further group of six animals matched for age, sex, and body weight served as controls and were given access to the pellet diet during the fasting period. At the end of this period, the control animals had gained a small but significant amount of weight (about 4% of initial body weight) whereas the fasted rats had lost about 7% of their initial weight.

Tissue Preparation and Sampling

All animals were killed under anesthesia by intracardiac perfusion of 10 ml of physiological saline followed by 500 ml of 2.5% glutaraldehyde in 0.1 M phosphate buffer (pH 7.2–7.4) at body temperature. Each perfusion took about 40 min.

Following perfusion, the entire small intestine of each rat was removed and its length $L(i)$ from pylorus to ileocecal junction was recorded. In the first experiment, intestines were sliced transversely with a razor blade into five segments of roughly equal length $L(s)$ and the luminal contents were washed out with fixative solution. A sample of intestine approximately 2 cm long was taken from each segment and stored in the fixative for further processing. In the second experiment, 2-cm samples were removed from intestines sliced into just three segments.

All samples of tissue were postfixed in 1% phosphate-buffered osmium tetroxide and subsequently sliced transversely into smaller pieces. Thereafter they were dehydrated in serial concentrations of ethanol and the pieces from each sample were flat-embedded in Araldite to facilitate orientation. Pieces were then removed with a small hacksaw and reembedded on dummy Araldite blocks so that they could be sectioned transversely.

From the pool of blocks from one segment from one rat, a single block of tissue was selected by lottery. This was repeated for all segments in each animal. One arbitrary transverse section across the entire intestine was cut from every block. All semithin sections (thickness ca. 0.5 μm) so obtained were stained with toluidine blue. Where ultrathin (ca. 70 nm) sections were required, block faces were first trimmed until 30–45° sectors of intestine remained. These sections were stained with lead citrate and uranyl acetate.

Light and Electron Microscopy

One or more micrographs from each semithin section were recorded on a Vickers M17 microscope using an orange filter. All were printed at a final magnification of $\times 51$ (first experiment) or $\times 56$ (second experiment), as determined by calibration with a stage micrometer scale. Where necessary, photomontages of complete transverse sections were constructed from these prints.

Ultrathin sections were examined using a Philips 201C electron microscope operated at an accelerating voltage of 60 kV. With the bars of the supporting copper grids as a convenient reference, fields of view displaying sectional images of the surfaces of villi were recorded in a systematic random fashion. Micrographs were printed to a final magnification of $\times 17,800$. Print magnifications were calibrated with the aid of a carbon grating replica bearing 2160 lines/mm.

Morphometry and Stereology

On each light micrograph a simple quadratic test lattice of spacing 1 cm (equivalent to 0.198 mm on specimens at $\times 51$ and to 0.179 mm at $\times 56$) was independently positioned and orientated. Intersections were recorded separately whenever horizontal and vertical test lines met sectional images of the primary mucosal and villus surfaces. For the single transverse section taken from one segment, the intersection counts $I(p)$ and $I(v)$ were summed to obtain the segmental ratio $I_t(v, p)$. The villus surface amplification factor was then calculated according to formula (3) and the circumference of the primary mucosa as shown in formula (5). The latter estimate was multiplied by segment length to calculate the primary mucosal surface in centimeters squared using formula (6). This estimate subsequently provided a value of the total villus surface area for one segment, according to formula (1).

Values of $S(p)$ and $S(v)$ summed over the separate segments from one animal allowed us to estimate total surface areas per rat.

Intersection counts $I(m)$ and $I(v)$ were summed over electron micrographs using the same test lattice (spacing now equivalent to 0.562 μm on the specimen). Relevant structural quantities characterizing the microvilli were obtained from formulas (4) and (2). In this study, results for microvilli are confined to the most proximal segment of animals in the first experiment.

Data Handling

Initially the values of each variable were calculated per intestinal segment. Differences between segments within animals were tested for significance using a one-way analysis of variance (first experiment). In the case of fasted versus nonfasted animals (second experiment), a two-way analysis of variance was performed to explore the sources of variation due to segmental and fasting effects. The interaction values obtained by applying this test gave an indication of the dependence of fasting effects on the level of segmental effects.

Mean values from each rat within a particular experimental group were also employed to calculate group means together with their standard errors (SEM).

Descriptions of these statistical procedures may be found in Bailey[23] and in Bishop.[24] All data were handled by a Hewlett-Packard HP-85 personal computer.

RESULTS

Experiment One

Mean intestinal length in this group of normal rats was estimated to be 84 cm (SEM 3.0 cm) and this represents an average segment length of almost 17 cm. In general, heavier individuals tended to have longer small intestines.

Table 1 shows the relative and absolute mucosal surfaces at different locations. In Table 2, we provide the one-way analysis of variance results based on the estimated segmental surfaces. Figure 2 illustrates the changing intestinal architecture in passing from proximal to distal segments.

There were significant proximodistal gradients of villus amplification factors ($F = 45.2$; d.f. 4, 25; $P < 0.001$), primary mucosal surface areas ($F = 5.4$; $P < 0.01$), and total villus surface areas ($F = 6.1$; $P < 0.01$).

In proximal segments 1 and 2, which together accounted for 40% of total intestinal length, villus amplification was remarkably uniform at about 7.4 times. In succeeding segments, the amplification factor decreased progressively to a figure of only 2.8 times in segment 5.

Since segment length was approximately constant within a given animal, the increasing proximodistal primary mucosal surface must be interpreted in terms of a larger intestinal caliber. This was supported by estimates of the mucosal circumference, which increased from 10 mm (SEM 0.94 mm) in segment 1 to 13 mm (0.33 mm) in segment 5 (see also Fig. 2).

In contrast to the primary mucosal surface, the surface area due to villi tended to decrease toward the terminal ileum. The difference between proximal and distal segments was on the order of 50%.

In the entire intestine, the amplification due to villi was, on average, 5.32 times (SEM 0.17). This overall amplification factor was determined by a primary surface of 96 cm² (SEM 5.6 cm²) and a villus surface of 512 cm² (34.1 cm²).

The estimated group mean microvillus amplification factor in segment 1

TABLE 1. Relative and Absolute Mucosal Surface Areas at Various Sites in a Group of Six Normal Rats

Variable	Segment 1	Segment 2	Segment 3	Segment 4	Segment 5
1. $S_S(v, p)$					
Group mean	7.41	7.36	6.25	4.02	2.81
SEM	0.189	0.303	0.394	0.306	0.306
2. $S(p)$, cm²					
Group mean	17.1	15.0	19.6	22.9	21.9
SEM	2.12	0.63	1.59	1.29	0.93
3. $S(v)$, cm²					
Group mean	127	110	122	92	62
SEM	16.5	4.1	11.6	8.7	8.2

TABLE 2. One-Way Analysis of Variance Results for Segmental Estimates of $S_S(v, p)$, $S(p)$, and $S(v)$

Variable	Source of variation	Sum of squares	Degrees of freedom	Mean square	Variance ratio, F
1. $S_S(v, p)$	Segments	102.1	4	25.53	45.2*
	Residual	14.1	25	0.56	
	Total	116.2	29		
2. $S(p)$	Segments	256.6	4	64.1	5.38**
	Residual	298.2	25	11.9	
	Total	554.8	29		
3. $S(v)$	Segments	16,686	4	4172	6.14**
	Residual	16,985	25	679	
	Total	33,671	29		

* $P < 0.001$; ** $P < 0.01$.

was 15.0 times (SEM 0.42). These figures are based on analysis of three animals, for which the segmental microvillus surface area was estimated to be 1450 cm² (SEM 73.3 cm²).

Experiment Two

Mean intestinal length in the control group was 100 cm (SEM 3.5 cm), giving an average segment length of about 33 cm. Corresponding lengths in fasted animals were 91 cm (6.7 cm) and 30 cm.

Table 3 displays the surface data for nonfasted and fasted rats. In Table 4 we summarize the results of two-way analysis of variance performed on these surface data.

Segmental differences in villus amplification and primary mucosal surface area were confirmed in these groups. However, there was no significant segmental variation in the secondary (villus) surface area.

Fasting had significant effects on both the primary and secondary mucosal surfaces but no significant influence on surface amplification by villi. The interaction between segmental and fasting effects was not significant for any of these estimates.

In the control group, villi amplified the entire intestinal surface 6.13-fold (SEM 0.24) and this figure was not altered significantly by fasting (estimated group mean 5.81, SEM 0.12).

Control rats possessed, on average, 83 cm² (SEM 4.8 cm²) of primary and 510 cm² (42.8 cm²) of secondary mucosal surface. In contrast, fasted rats had only 69 cm² (3.6 cm²) and 403 cm² (26.2 cm²), respectively. Maintenance of overall villus amplification after fasting could therefore be attributed to near-equiproportional decreases in both mucosal surfaces.

DISCUSSION

Using a stereological methodology developed recently,[16] we have been able to demonstrate the existence of significant anatomical gradients of mucosal

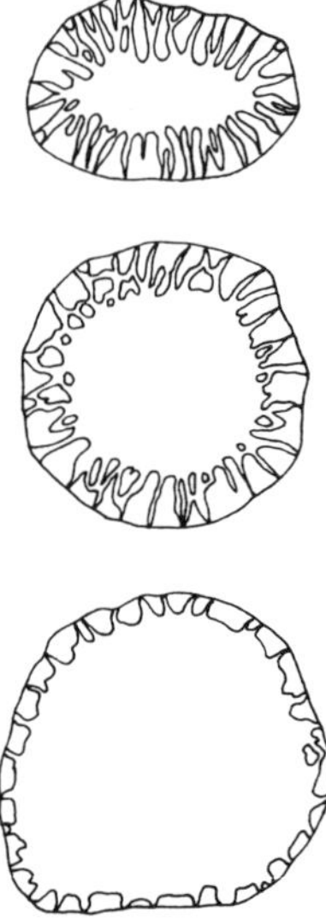

FIGURE 2. Tracings taken from semithin transverse sections through segments 1 (upper), 3 (middle), and 5 (lower) of a normal rat. Sectional images of the primary mucosa and of villi are shown. Notice the changing proximo-distal pattern of mucosal architecture and the distal increase in intestinal caliber.

architecture in the small intestines of normal rats and to detect significant alterations in this architecture in response to fasting. Two aspects of the present investigation are worthy of special comment: (1) the merits and limitations of the methodology and (2) the biological relevance of our findings.

Technical Considerations

The advantages of the present over previous methods have been discussed in more detail elsewhere.[16] The principal benefits may be summarized as follows:

1. They require no a priori assumptions about the sizes and shapes of villi. Therefore, they are superior to methods that rely on estimates of the dimensions of the ''average'' villus[2,6,7] and can be applied to study the

TABLE 3. Relative and Absolute Mucosal Surface Areas at Various Sites in Six Nonfasted and in Six Fasted Rats

Variable	Nonfasted			Fasted		
	Seg. 1	Seg. 2	Seg. 3	Seg. 1	Seg. 2	Seg. 3
1. $S_S(v, p)$						
Group mean	7.21	6.61	4.84	6.82	6.01	4.81
SEM	0.302	0.448	0.261	0.217	0.152	0.164
2. $S(p)$, cm^2						
Group mean	24.1	28.8	29.7	20.8	23.3	25.0
SEM	1.06	1.70	2.92	1.43	0.96	1.53
3. $S(v)$, cm^2						
Group mean	173	191	146	141	141	121
SEM	9.6	17.6	19.9	8.7	8.4	10.8

villus surface at different sites along the intestine as well as alterations in this surface that occur during aging, experimental manipulation, and disease. The new method of vertical sectioning[25] is also independent of size and shape but does not give the convenient visual impression of mucosal architecture afforded by transverse sectioning.

2. All estimates may be taken from transverse sections through the small intestine. This helps to improve the precision of estimation and to reduce the sampling workload. The time saved may be employed to analyze more segments from more animals.

3. The essential conditions of the underlying model (namely, that the primary mucosa is a cylinder and that villi display collective isotropy) probably hold for other species and for other tissue-processing procedures, e.g., immersion fixation and paraffin embedding. In any case, the validity of the model can be tested by making pilot studies on sections cut both transversely and longitudinally.

TABLE 4. Summary of Two-Way Analysis of Variance Performed on Surface Estimates

Variable	Variance ratios, F		
	Segments[a]	Fasting[b]	Interaction[a]
1. $S_S(v, p)$	32.7***	2.2 (NS)[c]	0.77 (NS)
2. $S(p)$	4.4*	10.2**	0.21 (NS)
3. $S(v)$	3.1 (NS)	10.8**	0.47 (NS)

[a] d.f. = 2,30.
[b] d.f. = 1,30.
[c] NS, not significant.
* $P < 0.05$; ** $P < 0.01$; *** $P < 0.001$.

Unfortunately, it is not yet established whether formula (3), and in particular the coefficient $4/\pi$, is applicable to biopsies of human material, since section orientation may be difficult to establish in such specimens. It is also pertinent to emphasize that the final estimates are "absolute" only in the sense of being internally consistent. When seeking to extrapolate surface areas from fixed and embedded to fresh unfixed tissues, it will be necessary to take into consideration such factors as the degrees of tissue shrinkage and/or distortion.[8]

Biological Implications

The villus amplification factor seems to be maintained over at least the proximal two-fifths of intestinal length and to diminish gradually thereafter. This gradient may be explained by disproportionate decreases in overall villus surface area since the extent of the primary mucosa actually *increases* in distal regions as the intestinal lumen becomes wider.

These results support observations made in other laboratories. Using alternative measures of villus surface amplification, several studies on the rat have noted that values tend to be higher proximally.[6,8,10] The regional differences correlate reasonably well with alterations in villus height,[1,2,5,8] although, being one-dimensional, this parameter is a less adequate indicator of surface changes.[8] On the basis of simplistic geometric models of villi, it has been suggested that the surface area of the average villus also declines toward the terminal ileum.[2]

Our estimates of the surface area due to villi in nonfasted rats are remarkably close to values of $510-530$ cm^2 for fixed and embedded intestines of length $88-115$ cm from rats of $170-325$ g body weight.[6,8] Moreover, villus amplification factors of $7.2-7.4$ in proximal segments are in excellent agreement with previous estimates of $6.8-7.3$ for the duodenum and proximal jejunum of slightly heavier Sprague-Dawley rats.[11] In the latter animals, the villus surface area of the duodenum can be estimated as 54 cm^2 and the microvillus surface area as 753 cm^2. Taking the present results and a mean duodenal length of 7 cm, we have calculated that our corresponding estimates would be 52 and 784 cm^2.

Variables of major adaptive and functional importance tend to be rigidly controlled. Moreover, low levels of interindividual variability are good indicators of the importance of these variables to the organism.[26,27] The external comparisons drawn above are strongly suggestive of minimal interanimal variation. Indeed, in this investigation we observed repeatedly that there are but small interanimal differences in surface area estimates. A convenient expression of the variability between animals within a given group is the coefficient of variation CV ($=$ standard deviation/group mean). For the three groups of rats in the present experiments, the observed CV values were only $5-10\%$ for overall villus amplification factors, $12-14\%$ for the surface area of the entire primary mucosa, and 16–21% for the total surface due to villi. Furthermore, these figures represent *overestimates* of the true variability between animals because they incorporate variations due to the relatively large intraanimal (i.e., intersegmental) differences. We must conclude, therefore, that relative and absolute mucosal surfaces are biologically relevant and informative quantities.

This conclusion is completely consistent with current knowledge of the functional properties, constancy, and regulation of mucosal architecture in the rat and other species. Thus, enzymes involved in the breakdown of oligopeptides and oligosaccharides[10] are embedded in the digestive-absorptive surface (microvillus glycocalyx) of enterocytes,[28] particularly those in more proximal regions. The total number of villi per intestine is known to be fixed and does not vary appreciably with age or with experimental treatment.[29,30] Epithelial villus transit times show a pattern of decrease from pylorus to ileocecal junction, the value in the duodenum of the adult rat being roughly twice that in the terminal ileum.[1,2]

In view of the constancy of villus number, it follows that the 20% loss of villus surface area after fasting may be attributed to a decrease in surface area of individual villi. This is supported by our two-way analysis of variance results, which indicate no significant interaction between segmental and fasting effects. In other words, fasting had no preferential influence on villi in any given segment. Recent investigations have suggested that villi maintain their height but become more slender during fasting.[15] This may be due, at least in part, to a hypoproliferative response to fasting, because epithelial cell cycle times increase during starvation and shorten again on refeeding.[3]

In view of the preceding considerations, we conclude that the present methods define structural quantities of relevance to future studies on the kinetics, biochemistry, physiology, development, and pathology of the small intestine.

REFERENCES

1. Altmann GG, Enesco M: Cell number as a measure of distribution and renewal of epithelial cells in the small intestine of growing and adult rats. Am J Anat 121:319–336, 1967.
2. Clarke RM: Mucosal architecture and epithelial cell production rate in the small intestine of the albino rat. J Anat 107:519–529, 1970.
3. Aldewachi HS, Wright NA, Appleton DR, Watson AJ: The effect of starvation and refeeding on cell population kinetics in the rat small bowel mucosa. J Anat 119:105–121, 1975.
4. Cheng H, Bjerknes M: Whole population cell kinetics of mouse duodenal, jejunal, ileal and colonic epithelia as determined by radioautography and flow cytometry. Anat Rec 203:251–264, 1982.
5. Lipscomb HL, Sharp JG: Effects of reduced food intake on morphometry and cell production in the small intestine of the rat. Virchows Arch B Cell Pathol 41:285–292, 1982.
6. Fisher RB, Parsons DS: The gradient of mucosal surface area in the small intestine of the rat. J Anat 84:272–282, 1950.
7. Lorenz-Meyer H, Köhn R, Riecken EO: Vergleich verschiedener morphometrischer Meth-

oden zur Erfassung der Schleimhautoberfläche des Rattendünndarms und deren Beziehung zur Funktion. Histochemistry 49:123–129, 1976.

8. Hromádková V, Skála I: Factors influencing the assessment of size of the mucosal surface and length of the small intestine in rats. Digestion 1:149–158, 1968.

9. Meinhard EA, Wadbrook DG, Risdon RA: Computer card morphometry of jejunal biopsies in childhood coeliac disease. J Clin Pathol 28:85–93, 1975.

10. Bastie MJ, Balas D, Laval J, Senegas-Balas F, Bertrand C, Frexinos J, Ribet A: Histological variations of jejunal and ileal mucosa on days 8 and 15 after hypophysectomy in rat: morphometrical analysis in light and electron microscopy. Acta Anat 112:321–337, 1982.

11. Stenling R, Helander HF: Stereological studies on the small intestinal epithelium of the rat. 1. The absorptive cells of the normal duodenum and jejunum. Cell Tissue Res 217:11–21, 1981.

12. Hall GA, Parsons KR, Batt RM, Bunch KJ: Quantitation of small intestinal structure and function in unthrifty piglets. Res Vet Sci 34:167–172, 1983.

13. Kessel RG, Kardon RH: Tissues and Organs: A Text-Atlas of Scanning Electron Microscopy. San Francisco: Freeman, 1979.

14. Baker SJ, Mathan VI, Cherian V: The nature of the villi in the small intestine of the rat. Lancet i:820, 1963.

15. Ross GA, Mayhew TM: Effects of fasting on villus morphology in different regions of the small intestine of the rat. J Anat 137:805–806, 1983.

16. Mayhew TM: Geometric model of the rat intestinal mucosa for stereological evaluation of villus amplification factors. J Microsc 135:337–346, 1984.

17. Mayhew TM: Basic stereological relationships for quantitative microscopical anatomy—a simple systematic approach. J Anat 129:95–105, 1979.

18. Weibel ER: Stereological Methods, volume 1, Practical Methods for Biological Morphometry. London: Academic Press, 1979.

19. Mayhew TM: On the relative efficiencies of alternative ratio estimators for morphometric analysis of cell membrane surface features. J Microsc 122:7–14, 1981.

20. Paumgartner D, Losa G, Weibel ER: Resolution effect on the stereological estimation of surface and volume and its interpretation in terms of fractal dimensions. J Microsc 121:51–63, 1981.

21. Gundersen HJG: Estimation of tubule or cylinder L_V, S_V and V_V on thick sections. J Microsc 117:333–345, 1979.

22. Weibel ER, Paumgartner D: Integrated stereological and biochemical studies on hepatocytic membranes. II. Correction of section thickness effect on volume and surface density estimates. J Cell Biol 77:584–597, 1978.

23. Bailey NTJ: Statistical Methods in Biology. London: English Universities Press, 1972.

24. Bishop ON: Statistics for Biology, 2d ed. London: Longman, 1971.

25. Baddeley AJ, Gundersen HJG, Cruz-Orive LM: Estimation of surface area from vertical sections. J Microsc 142:259–276, 1985.

26. Gundersen HJG, Seefeldt T, Østerby R: Glomerular epithelial foot processes in normal man and rats. Distribution of true width and its intra- and interindividual variation. Cell Tissue Res 205:147–155, 1980

27. Gupta M, Mayhew TM, Bedi KS, Sharma AK, White FH: Interanimal variation and its influence on the overall precision of morphometric estimates based on nested sampling designs. J Microsc 131:147–154, 1983.

28. Waldeck F: Functions of the gastrointestinal canal. In: Human Physiology, edited by Schmidt RF, Thews G, pp. 587–609. Berlin: Springer-Verlag, 1983.

29. Forrester JM: The number of villi in rat's jejunum and ileum: effect of normal growth, partial enterectomy and tube feeding. J Anat 111:283–291, 1972.

30. Clarke RM: The effect of growth and of fasting on the number of villi and crypts in the small intestine of the albino rat. J Anat 112:27–33, 1972.

chapter 9

practical ways to correct cytomembrane surface densities for the loss of membrane images that results from oblique sectioning

T. M. Mayhew and A. Reith

INTRODUCTION

Even experienced electron microscopists employ different criteria for recognizing images of cytomembranes in ultrathin sections. Consequently, they use different counting (or measuring) conventions for quantifying them.

While partly subjective, the ability to identify ultrastructural components is influenced strongly by technical considerations such as specimen contrast, resolution, section thickness, and sectioning angle. The interplay of these influences determines whether raw estimates of membrane surface density are systematically overestimated or underestimated. It is therefore advisable to scrutinize carefully the conditions and conventions that obtain in a given experimental situation so that appropriate correction procedures can be applied and more reasonable comparisons can be drawn between estimates made in different laboratories.

When considering these systematic errors, it is convenient to recognize two main categories of cytomembrane.[1] Class A membranes are associated with organelles for which the presence of a limiting membrane can be presumed even in regions where its image cannot be observed on the section. This is made possible by the changes in appearance between the organelle interior and the surrounding cytoplasm.[2–5] Outer mitochondrial membranes, the membranes of the nuclear envelope, and the membranes of various types

We thank Dr. Werner Hax and his colleagues at N. V. Philips Gloeilampenfabrieken (Eindhoven) for kindly undertaking the tilting stage micrography on our behalf. We are also grateful to Dr. G. H. Cope (Sheffield) for providing us with micrographs of parotid acinar cells. This collaboration was assisted by a travel grant from the British Council Academic Links and Interchange Scheme.

of inclusion granule fall into this first category. Such membranes seem to be less sensitive to the errors mentioned above in that there is relatively good accord between surface density estimates obtained by different observers or by the same observer working at different magnifications.[1,6]

Class B membranes, on the other hand, characterize organelles and organelle subcompartments that are not always so conveniently predictable. Into this class fall the membranes of mitochondrial cristae and rough and smooth endoplasmic reticulum. These membranes are more susceptible to systematic errors insofar as published estimates of their surface densities show larger discrepancies.[1,6]

Practical methods of correcting for section thickness (or overprojection) effects and for "resolution" effects are available.[6,7] To correct for membrane loss by oblique sectioning,[8] a factor of 1.5 has been used for cristae and endoplasmic reticulum.[2,9,10] However, this correction factor is not universally appropriate. There is evidence that it is unsuitable for other types of cytomembrane, other counting conventions, and other preparative conditions.[5,11-14]

In this chapter we describe and compare two methods of correcting for loss of membrane images due to oblique sectioning. In the first method, correction factors for class A membranes are calculated from the observed fractional losses of images. In the second, factors for class A and class B membranes are derived by goniometry. The two approaches are flexible enough to suit the conventions and conditions that prevail in other laboratories.

THE PROBLEM OF OBLIQUE SECTIONING

We begin by reviewing the theoretical background of this problem insofar as it relates to the aims of the present investigation.

Theoretical Foundation

Biological membranes examined by transmission electron microscopy may be tilted at various angles to the electron axis. These angles are determined by those at which the section intersects the perpendiculars to the membranes. At small angles the "unit membrane" structure can be observed clearly, but at larger angles this arrangement is no longer discernible, and the membrane images become vague and disappear rather abruptly.[8,9,12,14] As a consequence of this phenomenon, membrane surface densities will be underestimated. The degree of error depends partly on the convention adopted, i.e., on whether "clear" or "clear + vague" images are quantified.

To compensate for image loss, a correction factor of 1.5 has been proposed, this value being based on studies of the nuclear membranes of hepatocytes.[9] Roughly 50% of these membranes were found to disappear when tilted at an angle of 58° and some 90% at an angle of 70°. Therefore, it was suggested that for membranes for which oblique sectioning from 0 to 90° is equally likely, an angle of 58° would result in roughly two-thirds of membrane images being visible.

The validity or otherwise of extending the factor 1.5 to other species of membrane depends crucially on the critical angle θ beyond which the individual observer regards membrane images as being "lost" and so excluded from quantification. The smaller this angle, the larger the correction factor that is required. If only "clear" images are evaluated, the angle will be smaller than if "clear + vague" images are included.

The fundamental stereological relationship for estimating the surface density S_V of membranes in a reference volume is

$$S_V = CB_A$$

where B_A is the mean boundary trace length of membrane images on sections of reference area A and C is a numerical coefficient determined by the angle of sectioning and the degree of anisotropy (preferential orientation) of the membranes. When section thickness is zero and the sections are isotropic uniform random sections hitting the specimen, the coefficient C is equal to $4/\pi$ (see, for example, Weibel,[15] part 2.3.3., p. 36). The formula

$$S_V = \frac{4}{\pi} B_A$$

can then be invoked for arbitrary surfaces cut by independent isotropic uniform random sections or for isotropic surfaces cut by arbitrary sections. However, the formulation is strictly valid only if all possible membrane images can be seen on the sections. If there is a critical angle θ $(0 \leqslant \theta \leqslant \pi/2)$ above which the observer considers images to be lost, the true membrane surface density should be estimated instead using the relation

$$S_V = \frac{4}{\sin 2\theta + 2\theta} B_A$$

as shown elsewhere.[12,14] It follows that the raw or uncorrected surface density estimate must be increased by a factor K determined by the angle θ, where

$$K = \frac{\pi}{\sin 2\theta + 2\theta} \tag{1}$$

Class A Membranes

If the membrane belongs to class A, it is not necessary to estimate θ in order to calculate K. Instead, a correction factor can be derived from the observed overall fractional loss of membrane images. In practice, it is preferable to do this by intersection counting[15] rather than by trying to measure boundary trace lengths.

To illustrate the approach, let I_c denote the total number of test line intersections with the observed membrane images (these may be "clear" or "clear + vague," depending on the convention). Now let I_t represent the number of intersections with the entire organelle boundary (whether its membrane images are "clear," "vague," or "invisible"). Then the fractional loss of images, f, for any particular convention can be approximated by

$$f = \frac{I_t - I_c}{I_t} \tag{2}$$

and the corresponding correction factor will be

$$K = \frac{1}{1 - f} \tag{3}$$

Class B Membranes

With membranes of this class, it is not possible to estimate f by the above approach because (1) there is no visual means of compensating for membranes that are "invisible" and (2) "vague" images cannot always be identified unequivocally as belonging to the organelle concerned (e.g., endoplasmic reticulum). These empirical observations doubtless explain why formula (3) has been applied to *outer mitochondrial membranes*[3,4] in attempts to find a value of K for correcting raw surface density estimates of *cristae membranes*. The working assumption is that K for a class A membrane is the *same* as that for a class B membrane.

Since K cannot be estimated for class B membranes by intersection counting, we are forced to evaluate the relevant angle θ and to substitute pertinent values into formula (1). The angle θ may be estimated directly by goniometry. This approach has the advantage of allowing us to check whether or not K is the same for different types of membrane. It also allows us to compare goniometric with fractional image loss estimates.

From formulas (1) and (3), we see that θ may be estimated for class A membranes using

$$f = 1 - \frac{\sin 2\theta + 2\theta}{\pi} \tag{4}$$

Table 1 illustrates the theoretical relationships between f, θ, and K for iso-

TABLE 1. Theoretical Values of Fractional Loss and Correction Factor for Various Angles of Membrane Tilt

Angle θ (deg)	Fractional image loss f	Correction factor K
0	1	∞
5	0.889	9.03
10	0.780	4.55
15	0.674	3.07
20	0.573	2.34
25	0.478	1.92
30	0.391	1.64
35	0.312	1.45
40	0.242	1.32
45	0.182	1.22
50	0.131	1.15
55	0.090	1.10
60	0.058	1.06
65	0.034	1.04
70	0.018	1.02
75	0.008	1.01
80	0.002	1.00
85	0.000	1.00
90	0	1

tropic uniform random sectioning of arbitrary membrane surfaces or for arbitrary sectioning of isotropic surfaces (see also Mall et al.[12]).

In this study, formulas (1)–(4) are applied to goniometric and fractional image loss data in order to estimate f, θ, and K for different species of cytomembrane taken to represent the classes A and B.

MATERIALS AND METHODS

For present purposes, estimates are based on the convention of quantifying only "clear" membrane images, a practice that we adopted in an earlier investigation.[10] Electron micrographs illustrating this convention, as applied to cristae membranes, are given elsewhere.[1]

The Fractional Loss Approach

As explained above, this approach is suitable only for class A membranes. As examples of this class, we have chosen outer mitochondrial membranes (in rat hepatocytes and mononuclear phagocytes) and inner nuclear membranes (in rabbit parotid acinar cells).

Outer Mitochondrial Membranes

Rat livers were processed and sampled by procedures described elsewhere.[10] Silver to gray sections were stained with lead citrate. Electron micrographs of hepatocytes in midzonal regions of lobules from one liver (left lobe) were obtained by systematic sampling procedures and printed to a final magnification of $\times 29{,}300$. A total of 47 micrographs were employed for intersection counting. To this end, a multipurpose test system bearing 50 test lines, each of a length equivalent to 0.48 μm on the actual specimen, was positioned independently on each micrograph.

We counted intersections between test lines and outer mitochondrial membranes, I_c being the number of intersections with "clear" images and I_t the number of intersections with mitochondrial margins, whether "clear," "vague," or "invisible." The fractional membrane image loss was calculated from formula (2) after summing intersection counts over all micrographs. The standard error of the mean (SEM) for the set of micrographs was calculated using the equation given by Cochran (Ref. 16, part 2.11, p. 32). An estimate of K was then obtained using formula (3) and a value of Θ using formula (4).

To test for any resolution effect,[6] the above proceedings were repeated on an independent set of 25 micrographs printed to a final magnification of $\times 100{,}000$.

Pellets of mononuclear phagocytes from the peritoneal cavity of rats were obtained by methods described by Mayhew and Williams.[17] Sections of silver interference color were stained with lead citrate. Systematic random samples of cell profiles were printed to a final magnification of $\times 28{,}200$. For intersection counts with mitochondrial profiles, a single quadratic test lattice (spacing equivalent to 0.36 μm on the specimen) was superimposed independently on each of 18 micrographs. Intersections I_c and I_t with outer membranes were summed over "horizontal" and "vertical" test lines over all micrographs. Estimates of f, SEM, K, and θ were computed as described already.

Inner Nuclear Membranes

Rabbit parotid glands were processed and sampled by procedures detailed in Cope and Williams.[18] Ultrathin sections (silver interference color) were stained with methanolic uranyl acetate and lead citrate. Systematic random samples of micrographs of acinar cells from one gland of one animal were recorded and printed at a final magnification of $\times 34{,}000$. For intersection counting, a quadratic lattice of spacing 0.29 μm was superimposed independently on each of 40 micrographs displaying nuclear profiles. Other steps were as described for mitochondrial membranes in mononuclear phagocytes.

The Goniometric Approach

This approach was applied to membranes of classes A and B. As examples of class A, we selected outer mitochondrial and plasma membranes; for class B, we examined cristae mitochondriales. All membranes were from rat liver or kidney.

Tissues were fixed in *sym*-collidine-buffered osmium tetroxide and embedded in Epon. Ultrathin sections (silver to gray) were stained with lead citrate and examined on a goniometer stage in a Philips EM 400T electron microscope operated at an accelerating voltage of 60 kV.

Micrographs of liver were taken in midzonal regions of lobules to maintain consistency with fractional image loss analyses. All were printed at a nominal magnification of ×40,000. Kidney micrographs depicted epithelial cells of cortical convoluted tubules and were printed at a nominal ×25,000.

Micrographs in each series were taken at varying angles of specimen tilt, from $-42°$ to $+42°$ around the $0°$ tilt position. The angular increment was $6°$. Subsequent assessments were made on "clear" membrane images that were roughly orthogonal to the surface of the section (i.e., parallel to the axis of tilt) at $0°$ of specimen tilt. Membrane images were followed on both sides of this axis to the points at which they could no longer be regarded as "clear." The total range of tilt embracing "clear" images therefore corresponded to an angle of 2θ.

Altogether we made 10 observations on each of the following membranes: outer mitochondrial (liver), cristae (liver), outer mitochondrial (kidney), and plasma membrane (kidney). A further set of 12 observations was made on the cristae of kidney mitochondria.

For each set of observations on a given membrane, we calculated the mean value of θ together with the SEM for that set. Estimates of K were calculated using formula (1).

RESULTS

Our estimates of angles θ and correction factors K are summarized in Table 2. Values of θ are given to the nearest $1°$.

Fractional Loss Studies

From the sample of hepatocytes printed at ×29,300, the overall fractional image loss of outer mitochondrial membranes amounted to 0.570 (SEM 0.017). The corresponding correction factor was 2.33, which is equivalent to an angle of about $20°$ by formula (4).

Repeating the analyses on micrographs printed at ×100,000 yielded very similar values: $f = 0.591$ (SEM 0.032), $K = 2.44$, and $\theta = 19°$.

For the outer mitochondrial membranes of mononuclear phagocytes, we obtained $f = 0.558$ (SEM 0.028), $K = 2.26$, and $\theta = 21°$.

For the inner nuclear membranes of parotid acinar cells, the estimated loss was 0.555 (SEM 0.017). This degree of image loss represents a correction factor of 2.25 and a critical angle of $21°$.

Goniometric Studies

The outer mitochondrial membranes of hepatocytes were no longer "clear" beyond an angle of $21.2°$ (SEM $2.7°$). Using formula (1), the appro-

TABLE 2. Values of θ and K Estimated for Different Types of Cytomembrane

Membrane class and species	Estimated from fraction of images "lost"[a]		Estimated by goniometry	
	θ (deg)	K	θ (deg)	K
Class A				
Outer mitochondrial				
Liver[b]	19–20	2.33–2.44		
Phagocytes	21	2.26		
Liver[c]	20	2.32	21	2.22
Kidney[c]	20	2.34	22	2.13
Inner nuclear				
Parotid gland	21	2.25		
Plasmalemma				
Kidney			24	1.97
Class B				
Cristae mitochondriales				
Liver			26	1.85
Kidney			26	1.84

[a] This approach practicable for class A membranes only.
[b] Estimated from fractional losses obtained at two different magnifications.
[c] Fractional loss estimates based on one micrograph, at 0° of specimen tilt, from the complete goniometric series.

priate correction factor was calculated to be 2.22. (Incidental note: intersection counts performed on the micrograph at 0° of specimen tilt gave $f = 0.569$, $\theta = 20°$, and $K = 2.32$ for these membranes.)

Examining mitochondrial cristae in hepatocytes produced an estimated angle of 26.0° (SEM 2.5°) and a correction factor of 1.85.

In the kidney samples, outer mitochondrial membranes yielded an angle of 22.2° (SEM 1.0°), for which the required K would be 2.13. (Incidental note: intersection counts on the micrograph at 0° of tilt gave $f = 0.573$, $\theta = 20°$, and $K = 2.34$).

For cristae within these mitochondria, the estimates were 26.2° (SEM 1.3°) and 1.84. For plasma membranes, we obtained 24.3° (SEM 1.3°) and a correction factor of 1.97.

All goniometric analyses were undertaken by the same person. Subsequent examination of the same preparations by the coauthor produced essentially similar estimates: $\theta = 27°$ (outer mitochondrial membranes, liver), 24° (cristae, liver), and 24° (cristae, kidney).

DISCUSSION

The purpose of this investigation was to estimate factors that correct cytomembrane surface densities for image loss due to section obliquity. Such factors have been obtained by (1) intersection counting to estimate fractional image losses for a predefined counting convention and (2) goniometry to estimate the critical angle to which membranes must be tilted for their images to be excluded on the same convention. These methods have been applied to membranes belonging to two distinct classes (A and B) to test whether the same correction factor applies to different types of cytomembrane in the same preparations.

The use of the term "clear" in the present studies warrants further comment. The practice of quantifying only "clear" images in stained sections seems to correspond to one that identifies the unit membrane structure. Thus, in a goniometric analysis of the plasma membranes of muscle blood capillaries,[14] it was noted that the unit membrane structure could be observed in lead citrate-stained sections up to a tilt angle of 20°. Fractional image loss analysis of the same membranes suggested an angle of 26°. These figures agree well with present estimates. However, since the angle θ also depends on the contrast between membranes and the surrounding cellular components, the use of en bloc staining with uranyl acetate to enhance specimen contrast could conceivably alter the counting practice. This might be expected to increase the estimate of θ and reduce the required correction factor.

Our results indicate that the correction factor for dealing with membrane loss lies somewhere in the range 1.84–2.44 (Table 2) when "clear" images only are quantified. Goniometric estimates for outer mitochondrial (class A) membranes by both of the present authors were 1.80–2.22 and compared favorably with factors of 1.84–1.97 for cristae (class B) membranes in the same sections. These findings suggest that the practice of using factors determined for outer mitochondrial membranes to correct surface densities of cristae[3,4,12] may indeed be a satisfactory approach.

However, it must be emphasized that the practice should be confined to studies performed under standardized preparative conditions. Moreover, correction factors derived from goniometrically measured angles suppose isotropic uniform random sectioning and so the practice may not be valid when arbitrary sections are taken through membranes that differ in their orientation characteristics. Myocardial mitochondria, for example, show preferred directions of orientation and correction factors for outer mitochondrial membranes may not apply to cristae.[12] We also noticed that factors much larger than those indicated here may be required for outer mitochondrial membranes in anisotropic tissues cut in preferred directions. In specimens of the human nasal mucosa sectioned perpendicular to the epithelial surface, correction factors calculated by the fractional image loss method were found to be as great as 10.8 (A. Reith, unpublished observations).

In the present studies, correction factors varied minimally with membrane species; goniometric estimates of 1.97 for plasma membranes were very close to figures of 1.80–2.22 for mitochondrial membranes, and fractional loss estimates of 2.25 for nuclear membranes were close to values of 2.26–2.44 for outer mitochondrial membranes. Again, when the encounter between section and membrane is not isotropic uniform random, it will be advisable to monitor the orientation characteristics of different cytomembranes before transferring correction factors from one membrane species to another.

Comparison of present estimates with those obtained on other experimental systems and in other laboratories reveals some discrepancies. The correction factors of 2.26–2.44 for outer mitochondrial membranes in liver, kidney, and peritoneal cells are within the range 1.9–2.6 that can be calculated from fractional image loss analyses of these membranes in myocardiocytes from rabbits and rats.[3,12] However, our estimates are generally higher than the values of 1.16–1.5 advocated for cristae and endoplasmic reticulum membranes.[2,5,9,11] These differences are not surprising when one considers that estimates depend on the interplay of several influences, on the use of alternative counting conventions and variable technical conditions. These influences must be expected to vary between different observers and different laboratories.

The corrections proposed in this chapter account only for image loss due to oblique sectioning. However, this is but one of several sorts of technical bias that affect estimates of membrane surface densities. Loss of images of organelles can take other forms; for instance, profiles of organelles may lack sufficient contrast or it may not be possible to distinguish them unambiguously from similar sections through different organelles. Thus, near-tangential sections through hepatocyte mitochondria can render their profiles indistinguishable from profiles of similarly sectioned microbodies.[19] In the past, this problem of profile "truncation" or "capping" proved difficult to solve. Though procedures have been developed for certain situations,[20] the combined use of goniometry and fractional image loss analysis could offer an approach to estimating the influence of truncation on surface density data. For the present, it is advisable to note that the fractional image loss and goniometric methods may differ in susceptibility to truncation errors. In general, then, goniometry may be a safer way to determine correction factors for image loss due to section obliquity.

Among other biases for which corrections are available we can mention overprojection due to positive section thickness[7] and resolution.[6] Applying these corrections to suit the conventions and conditions prevalent in individual laboratories may go some way toward minimizing the large discrepancies between surface densities estimated by various research groups. For example, raw surface densities of cristae in the chondriome of normal rat hepatocytes have been estimated as 6.3 m^2/cm^3 at the one extreme[10] and 24.4 m^2/cm^3 at the other.[11] The lower figure was based on "clear" images in stained sections. Correcting for membrane image loss with a factor of 1.85 (Table 2) would raise this figure to 11.7 m^2/cm^3. However, Weibel and Paumgartner[7] have argued that their raw estimates should be corrected for overprojection using a correction factor of 0.76. This would bring the higher figure down to 18.5 m^2/cm^3.

It is tempting to speculate that some of the residual discrepancy is attributable to resolution effects. The estimate of 11.7 m^2/cm^3 comes from intersection counts performed at a magnification of ×40,000 and that of 18.5 m^2/cm^3 from counts at ×96,000. It is interesting that the figures are close to estimates of 9.2 and 16.6 m^2/cm^3 for cristae analyzed at respective magnifications of ×38,000 and ×99,000 (see Paumgartner et al.,[6] Table 1, p. 52). Unfortunately, it is not yet possible to isolate resolution effects from other influences since these authors[6] were not assessing true resolution—discrimination of two points—but also the observer's ability "to recognise 'faint' profiles." It is not clear, for instance, to what extent their observation that resolution had a minimal influence on the surface densities of outer mitochondrial membranes merely reflects the fact that these are class A membranes whose images are easier to "recognise." We have also found that correction factors for these membranes are but marginally affected by varying the magnification.

The situation is complicated further by the fact that counting "clear" membrane images does not preclude overestimation by section thickness effects. Nor will contrast enhancement by en bloc staining, coupled with counting "vague" or "faint" images, entirely eliminate underestimation due to oblique sectioning.

From these considerations, it seems preferable to estimate surface densities using a critical magnification in order to minimize possible resolution effects.[6] Thereafter, the choice of correction factor should be decided after considering the counting convention and preparative conditions. It may be necessary to combine corrections for oblique sectioning with those for overprojection.[7] The choice of method for dealing with image loss should also take into account the possible effects of truncation[20] as well as the orientation properties of the membranes and/or sections. In any event, it is strongly recommended that correction factors be estimated rather than relying on values employed by other laboratories.

On many occasions, "true" surface densities are not required, only values for comparing control and experimental groups. In these circumstances it may be enough to satisfy oneself that stereological estimates are internally consistent, but it is still preferable to test whether or not separate experimental groups are influenced differentially by systematic errors.

REFERENCES

1. Reith A, Mayhew TM: Caveat bei der morphometrischstereologischen Bestimmung von Biomembranen. Gegenbaurs Morphol Jahrb 126:206–215, 1980.
2. Loud AV: A quantitative stereological description of the ultrastructure of normal rat liver parenchymal cells. J Cell Biol 37:27–46, 1968.

3. McCallister LP, Page E: Effects of thyroxin on ultrastructure of rat myocardial cells: a stereological study. J Ultrastruct Res 42:136–155, 1973.
4. Reith A: Caveat on the estimation of surface density (SD) of biomembranes (BM): indication for a systematic error. Microsc Acta Suppl 1:205–207, 1977.
5. Jakovcic S, Swift HH, Gross NJ, Rabinowitz M: Biochemical and stereological analysis of rat liver mitochondria in different thyroid states. J Cell Biol 77:887–901, 1978.
6. Paumgartner D, Losa G, Weibel ER: Resolution effect on the stereological estimation of surface and volume and its interpretation in terms of fractal dimensions. J Microsc 121:51–63, 1981.
7. Weibel ER, Paumgartner D: Integrated stereological and biochemical studies on hepatocytic membranes. II. Correction of section thickness effect on volume and surface density estimates. J Cell Biol 77:584–597, 1978.
8. Williams RC, Kallman F: Interpretations of electron micrographs of single and serial sections. J Biophys Biochem Cytol 1:301–310, 1955.
9. Loud AV: Quantitative estimation of the loss of membrane images resulting from oblique sectioning of biological membranes. In: Proceedings 25th Anniversary Meeting of the Electron Microscopy Society of America, edited by Arceneaux CJ, pp. 144–145. Baton Rouge, La.: Claitor's Book Store, 1967.
10. Reith A: The influence of triiodothyronine and riboflavin deficiency on the rat liver with special reference to mitochondria. A morphologic, morphometric and cytochemical study by electron microscopy. Lab Invest 29:216–228, 1973.
11. Weibel ER, Stäubli W, Gnägi HR, Hess FA: Correlated morphometric and biochemical studies on the liver cell. I. Morphometric model, stereologic methods and normal morphometric data for rat liver. J Cell Biol 42:68–91, 1969.
12. Mall G, Kayser K, Rossner JA: The loss of membrane images from oblique sectioning of biological membranes and the availability of morphometric principles—demonstrated by the examination of heart muscle mitochondria. Mikroskopie 33:246–254, 1977.
13. Bolender RP, Paumgartner D, Losa G, Muellener D, Weibel ER: Integrated stereological and biochemical studies on hepatocytic membranes. I. Membrane recoveries in subcellular fractions. J Cell Biol 77:565–583, 1978.
14. Casley-Smith JR, Davy P: The estimation of distances between parallel membranes on thick sections. J Microsc 114:249–259, 1978.
15. Weibel ER: Stereological Methods, volume 1, Practical Methods for Biological Morphometry. London: Academic Press, 1979.
16. Cochran WG: Sampling Techniques, 3d ed. London: Wiley, 1977.
17. Mayhew TM, Williams MA: A quantitative morphological analysis of macrophage stimulation. I. A study of subcellular compartments and of the cell surface. Z Zellforsch Mikrosk Anat 147:567–588, 1974.
18. Cope GH, Williams MA: Improved preservation of parotid tissue for electron microscopy. A method permitting the collection of valid stereological data. J Cell Biol 60:292–297, 1974.
19. Berger ER: Two morphologically different mitochondrial populations in the rat hepatocyte as determined by quantitative three-dimensional electron microscopy. J Ultrastruct Res 45:303–327, 1973.
20. Cruz-Orive LM: Distribution-free estimation of sphere size distributions from slabs showing overprojection and truncation, with a review of previous methods. J Microsc 131:265–290, 1983.

part 4

length measurements

Blood vessels are essentially long branched tubes whose length density provides a convenient measure of tissue vascularity and metabolic requirements. Length density (L_v) is surprisingly easy to estimate from random and independent sections.

In this chapter by Østerby and Gundersen, the appropriate methods are illustrated by application to renal glomeruli and used to test for isomorphous growth of capillaries in diabetic renal hypertrophy. Essentially the same approach could be used on other biological tubules such as seminiferous tubules in the testis, neuronal cell processes, and various types of glandular duct.

chapter 10

stereological estimation of capillary length exemplified by changes in renal glomeruli in experimental diabetes

R. Østerby and H. J. G. Gundersen

INTRODUCTION

Patients with diabetes of long duration very often suffer from renal insufficiency due to diabetic glomerulopathy. This very serious aspect of diabetic microangiopathy develops as a slowly progressing basement membrane accumulation in the glomerular capillaries and may eventually lead to glomerular closure.

Quite another kind of renal structural alteration has been detected in recent years. Immediately after the onset of the abnormal diabetic metabolism, a marked enlargement of the whole kidney and of individual nephrons occurs. This holds in patients with juvenile diabetes mellitus[1,2] as well as in animals with a chemically induced diabetic state.[3,4]

In these early stages of diabetes, renal hyperfunction is also present.[1] In fact, the enlargement of the glomerular structures with the increase in capillary filtration surface[5] most likely forms the basis of the glomerular hyperfiltration state.[6]

Quite intriguing questions present themselves in this situation. What happens with the glomerular structures—the rather complicated capillary tangle—when the acute growth sets in? Since in experimental diabetes the glomerular volume increases by about 30% over the first 4-day period,[7] quite drastic structural alterations are occurring.

A simple structural quantity concerning capillary networks is the total cap-

The animal experiments were done in collaboration with K. Seyer-Hansen. The very skillful technical assistance of K. Gerlach, U. Dalsgaard Hansen, A. Larsen, and L. Nielsen is gratefully acknowledged. The study was supported by grants from the Danish Medical Research Council, the Danish Diabetic Association, P. Carl Petersen Foundation, Nordic Insulin Foundation, Novo's Foundation, and Århus University.

illary length. In the glomerular capillaries this parameter may well be of some functional relevance.[8] Our problem was then to estimate glomerular capillary length in the growing kidney in rats with induced diabetes.

STEREOLOGICAL ESTIMATION OF TUBULE LENGTH

The glomerular capillaries, although somewhat irregular in shape, may as a first approximation be considered as tubular structures, i.e., structures with an infinitely large length compared to the diameter.

The estimation of the length *density* L_V—that is, the ratio of tubular length to the containing volume—of such structures is very simple, requiring the counting of profiles on sections and estimation of the area representing the containing volume.

The stereologic formula relating the two dimensional appearance to the three-dimensional structure is

$$L_V = 2 \frac{\Sigma Q(\text{tub})}{\Sigma A(\text{ref})}$$

where $\Sigma Q(\text{tub})$ is the total number of tubular profiles counted within the total reference area, $\Sigma A(\text{ref})$, seen in one animal.[9]

From the length density, the total length is obtained by multiplying by the reference volume:

$$L(\text{tub}) = 2 \frac{\Sigma Q(\text{tub})}{\Sigma A(\text{ref})} V(\text{ref})$$

The one assumption that must be fulfilled for the relationship to hold is a random orientation between the tubular structures and the plane of sectioning.

In practice, we estimate glomerular capillary length density within the glomeruli by first obtaining an unbiased sample of sections through the glomeruli.[10] In a single glomerular cross section cut at an arbitrary level within the tuft, the requirement of random orientation between capillary axes and plane of sectioning is very likely to be fulfilled. Taking cross sections from two or three glomeruli, it definitely is so. It should be noted that the orientation problem in this case, as in general, must be handled in the process of cutting the sections.

The capillary profiles must be defined as either external profiles (i.e., the delineation of one profile is the epithelial aspect of the peripheral basement membrane) or internal profiles (i.e., capillary lumen delineated by the endothelial cell luminal cytoplasm). Two internal profiles may therefore be contained within one external profile. In other situations, however, the external profile does not contain an internal one since the capillary space in the plane of section is entirely covered by endothelial cytoplasm (Fig. 1). Within individual fields the number of external and internal profiles may not be the same, but the total number by either definition is identical. Counting external profiles leads to an estimate of the length of the external aspect of the capillary; internal ones correspond to the length of the inner side. The two, of course, are identical. In the glomeruli we have counted luminal profiles since isolated profiles of lumina are usually easier to identify than those of the external capillary wall.

In counting the capillary profiles we do not encounter any edge problem if we sample only glomerular cross sections that are complete. Capillary profiles can be identified reasonably well with low-magnification electron microscopy and we do not have to subdivide the cross section into smaller counting areas.

In other cases, when subfields must be counted, separately, the edge problem is solved by the (very simple) application of the unbiased counting frame.[11]

The second figure necessary for the estimation of L_V is the total reference area A, that is, the sum of the areas of the two or three glomerular profiles, estimated by simple point counting:

$$A(\text{glom}) = a_p \Sigma P(\text{glom})$$

where a_p is the area associated with one point, which in turn is the inverse of the point density in the test system: (number of points)/(area of test system). We used as our reference area the glomerular "string polygon" encircling the glomerular capillaries and thereby also containing some of the urinary space (Fig. 2). We did so because the corresponding area can be identified by light microscopy. At this level, glomerular volume as a fraction of the whole kidney, $V_V(\text{glom/kidney})$, is estimated by point counting on a series of systematically sampled sections obtained independently of the renal structures. If the volume of the kidney was determined beforehand, we now have all parameters necessary for estimating the absolute (i.e., not just the relative) structural quantity: the total length of glomerular capillaries in one kidney.

$L_V = 2Q_A(\text{cap/glom})$ estimates total capillary length per total glomerular volume. Multiplying by total glomerular volume $V(\text{glom})$ in the kidney, we obtain the capillary length in one kidney:

$$L(\text{cap}) = 2Q_A V(\text{glom})$$
$$= \left[\frac{2\Sigma Q(\text{cap})}{a_p \Sigma P(\text{glom})}\right] \frac{\Sigma P(\text{glom})}{\Sigma P(\text{kidney})} V(\text{kidney})$$

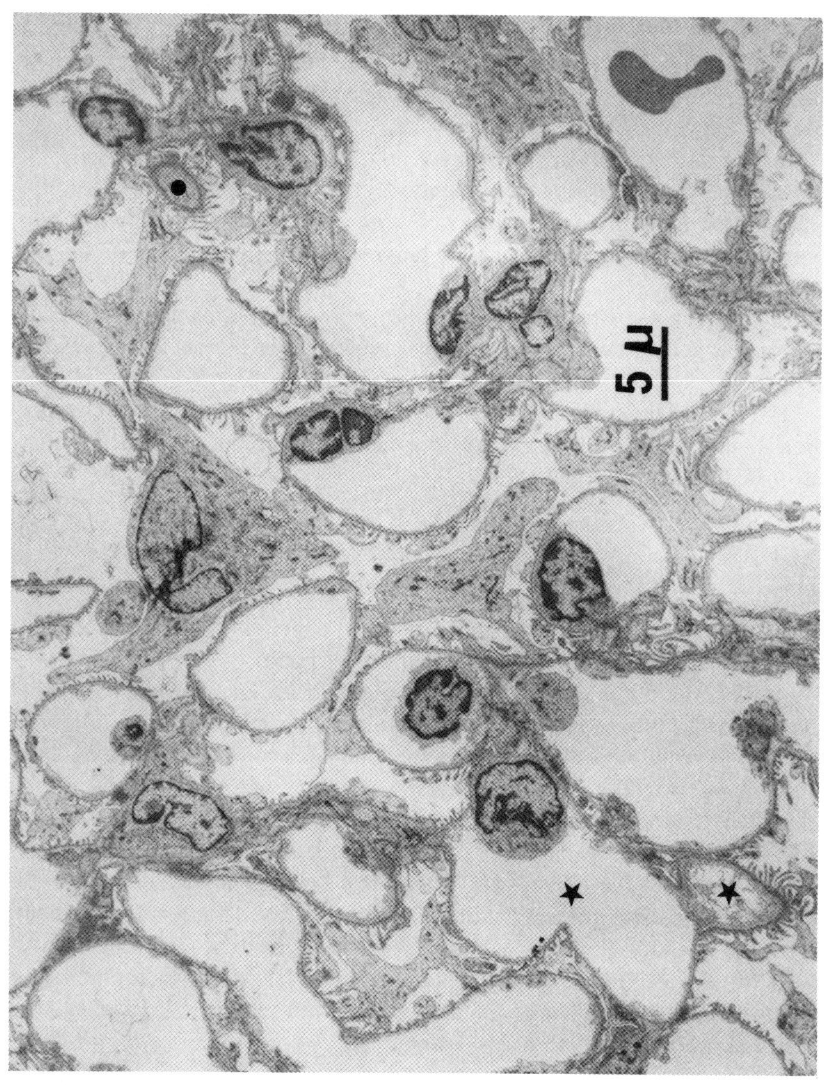

5 μ

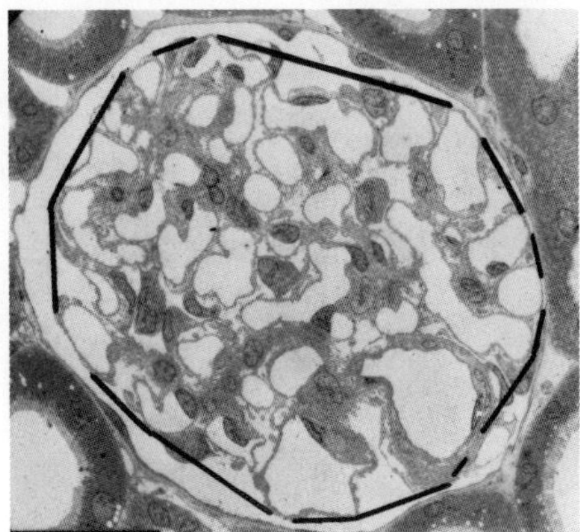

FIGURE 2. One whole glomerular cross section at low-magnification electron microscopy. The string polygon is the reference area common to light microscopic and electron microscopic levels. This identity of reference areas at two sequential levels is a prerequisite for the final estimate of absolute structural quantities pertaining to the whole organ (in this case one kidney).

BIASES IN TUBULE LENGTH ESTIMATES

Any systematic deviation of our estimates of tubule length from its true value is a bias in statistical terminology. Various factors may lead to such biases: the reality does not fit the model of infinitely long tubules, section thickness is not zero, preparation of the tissue for electron microscopy changes its dimensions, etc.

A number of the biases related to the geometric model are discussed in detail elsewhere.[12] Suffice it to state here that the tortuousity of the capillary axes and the deviation from circularity in the cross section do not give rise to any bias in the estimates. Moreover, the fact that the capillaries make up an interconnected network excludes the section thickness bias present in the estimation of the length of *isolated* tubules,[12] but it does give rise to another "geometric" bias. This is due to the fact that around a branching point, as

FIGURE 1. Section of a glomerular tuft from a perfusion-fixed rat kidney. For the estimation of capillary length either the internal (luminal) or the external profiles are counted. The stars show two luminal profiles separated by a bridge of endothelial cytoplasm. At this location only one external profile would be counted, whereas at the dot one external and no internal count would be recorded, since the inside is covered by endothelial cytoplasm.

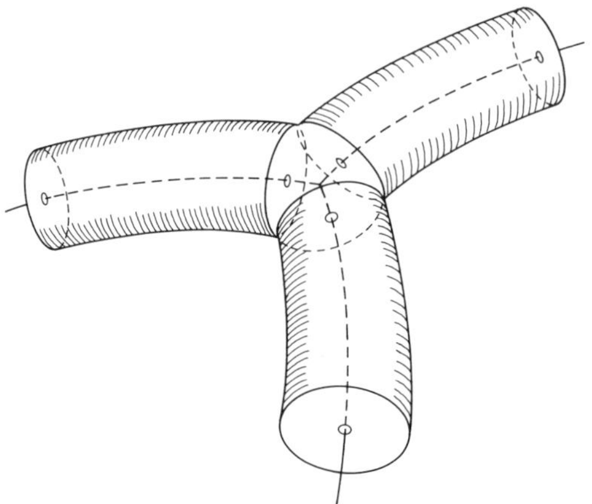

FIGURE 3. Sketch of a node or branching point where three tubules meet. The dashed curves are the three (imaginary) tubular axes whose total length is under study.

indicated in Fig. 3, there is a small "triangular" region of the tubular space where a section plane may generate only one profile while intersecting more than one of the tubular axes. The resulting *under*estimation of tubular length is of the order of d/l for the minimum of three tubular axes meeting in a point, where l is the average tubular length between two branching points. In general, l is not known since its estimation requires estimates of the number of branching points in the network. However, for comparative purposes we need not correct for the bias—at least as long as changes in d/l do not play a major role in observed changes in $L(cap)$.

To escape biases related to identification problems, we must use a magnification that permits easy recognition of the individual capillary profiles. This means that electron microscopy must be applied rather than high-magnification light microscopy. If we have the opportunity to use perfusion-fixed material, the identification can be made at a lower magnification than if immersion-fixed material must be used.

RESULTS IN EARLY
EXPERIMENTAL DIABETES

Rats injected with submaximal doses of streptozotocin develop a permanent moderate diabetic state. This animal model has been used extensively to study kidney lesions. Such rats develop the acute renal changes seen in diabetic patients as well as basement membrane thickening detectable after about 6 months.[13,14] These lesions correspond to the moderately advanced stages of diabetic glomerulopathy.

In this context we will consider only the early acute state. Immediately after the induction of diabetes, i.e., within the first 24-h period, the kidney starts to grow.[4] The glomeruli are markedly involved in this very early growth phase. After 4 days of diabetes the glomerular volume fraction is increased, which means that the glomeruli grow faster than the remaining parts of the kidney. We studied the alterations of the glomerular structures over the period from 4 to 47 days of diabetes.[7,8]

The results relevant to the present review are shown in Table 1. The presence of renal and glomerular hypertrophy after 4 days of diabetes is evident. It is also seen that the growth slows down after this initial burst. The total glomerular capillary length is found to be more than ½ kilometer in the control rats. After 4 days this length is increased by 32% ± 4 (SEM), whereas after 47 days it is again in the normal range. The capillary luminal cross-sectional area (obtained as the capillary volume divided by its length), on the other hand, is unchanged at 4 days but increased by 26% ± 6 after 47 days. [These findings show the very drastic modulations in capillary geometry taking place during the acute growth phase. The increase in capillary cross-sectional area and the decrease in L(cap) from day 4 to day 47 may both imply an increase in the "network bias" proportional to d/l. This means that we cannot be certain about the actual magnitude of the decrease in L(cap) at day 47.]

How do these quantitative alterations compare with the change in size of the whole glomerular tuft? Evidently, when the tuft enlarges we must necessarily observe some enlargement of the glomerular capillaries.

To compare the changes in length [of dimension (length)^1] with those in the glomerular volume [dimension (length)^3] we must compare equal dimensions by, e.g., raising the estimates of length to the power 3. If the glomerular capillaries show completely isomorphous growth, the ratio L^3/V will remain constant.

A plot of this "shape factor" in the three groups is shown in Fig. 4. These calculations can be done when we have obtained the total length and the total volume; they cannot be done on the length densities. It appears from Fig. 4 that the capillary length initially increases more than expected from the glo-

TABLE 1. Data on the Whole Left Kidney in Three Groups of Rats (Means ± SD)

Group	n	Kidney weight (mg)	Glomerular volume (mm³)	Glomerular capillary length (m)
Control rats	12	863 ± 84	42.6 ± 4.8	615 ± 50[a]
Diabetic rats after 4 days	12	1044 ± 87	56.0 ± 5.9	814 ± 59[a]
Diabetic rats after 47 days	8	1490 ± 159	61.1 ± 11.1	656 ± 112

[a] $n = 9$.

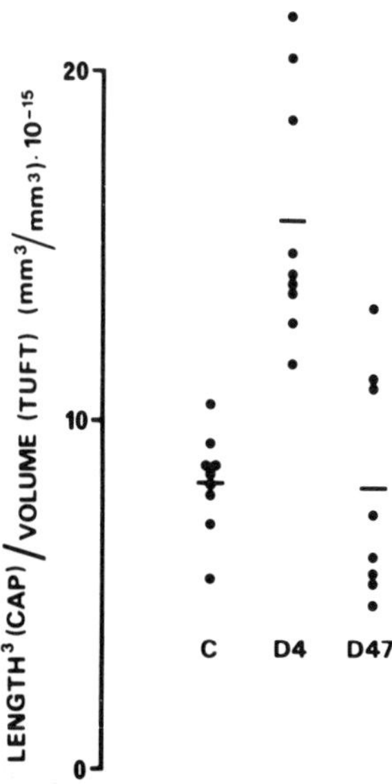

FIGURE 4. To compare changes in two different structural quantities the dimension must be the same. The plot illustrates the relationship between length and volume changes. In the D4 group (animals diabetic for 4 days) the increase in length (glomerular capillaries) is disproportionately larger than the concomitant increase in reference volume (glomerular hypertrophy). In group D47 the normal level is reestablished. In group D4 the glomerular composition is nonisomorphous; in group D47 it *may* be isomorphous.

merular expansion. After the 47-day period, however, remodeling of the capillaries seems to have taken place, with the structural composition of the tuft returning toward normal.

DISCUSSION

The example given in this chapter shows that it is very simple and easy to obtain estimates of the length of tubular structures. It is noteworthy that the structures need not be linear, so quite a large number of structures in biology may be quantified in this way. The assumptions are few and easy to fulfill, and the whole method is very robust.

As illustrated in the example, one should never restrict the observations to relative data (length density), since biologically meaningful interpretation of relative changes is usually not possible. One should also keep track of the dimensions when comparing various structural changes (see the example).

The main purpose of the present study was to compare various structural quantities in three different groups of animals. This means that we were not particularly concerned about the exact values of the structural quantities. In other situations, questions concerning the absolute dimensions may be the main issue, as when structural quantities are to be related to some physiological parameters. In that case we would have to make a thorough study of the effect of the preparative procedures on the tissue dimensions in terms of shrinkage, swelling, or possibly differential changes of individual compartments within the tissue. In our series we compared the specific gravities of kidney tissues in all groups and found them to be equal. This parameter enters when we apply kidney weight as the estimate of total reference volume. We also established that the increase in kidney weight due to the perfusion was the same in all groups. We are therefore confident in the comparison of structural quantities between the groups. But we do not know exactly how far off we are, e.g., concerning the figure for total capillary length in relation to the true, in vivo situation. Yet, obviously the estimate of 615-m glomerular capillaries in outgrown female rats does give a good impression of the order of magnitude of this structure.

The example given in this chapter shows that, if a proper sampling technique has been applied, the simple-to-perform stereological measurements lead to great insight into structural alterations in a given situation. Evidently, a semiquantitative—not to mention a qualitative—evaluation of the structures would not have provided much information about what happens to the glomerular capillaries during the acute growth phase immediately after the onset of a diabetic state.

REFERENCES

1. Mogensen CE, Østerby R, Gundersen HJG: Early functional and morphologic vascular renal consequences of the diabetic state. Diabetologia 17:71–76, 1979.
2. Østerby R, Gundersen HJG: Glomerular size and structure in diabetes mellitus. I. Early abnormalities. Diabetologia 11:225–229, 1975.
3. Ross J, Goldman JK: Effect of streptozotocin-induced diabetes on kidney weight and compensatory hypertrophy in the rat. Endocrinology 88:1079–1082, 1971.
4. Seyer-Hansen K: Renal hypertrophy in streptozotocin diabetic rats. Clin Sci Mol Med 51:551–555, 1976.
5. Kroustrup JP, Gundersen HJG, Østerby R: Glomerular size and structure in diabetes mellitus. III. Early enlargment of the capillary surface. Diabetologia 13:207–210, 1977.
6. Hirose K, Tsuchida H, Østerby R, Gundersen HJG: A strong correlation between glomerular filtration rate and filtration surface in diabetic kidney hyperfunction. Lab Invest 43:434–437, 1980.
7. Seyer-Hansen K, Hansen J, Gundersen HJG: Renal hypertrophy in experimental diabetes: a morphometric study. Diabetologia 18:501–505, 1980.

8. Østerby R, Gundersen HJG: Fast accumulation of basement membrane material and the rate of morphological changes in acute experimental diabetic glomerular hypertrophy. Diabetologia 18:493–500, 1980.

9. Weibel ER: Stereological Methods, volume 1, Practical Methods for Biological Morphometry. London: Academic Press, 1979.

10. Østerby R, Gundersen HJG: Samplings problems in the kidney. In: Geometrical Probability and Biological Structures: Buffon's 200th Anniversary, edited by Miles RE, Serra J, pp. 185–191. Lecture Notes in Biomathematics 23. Berlin: Springer, 1978.

11. Gundersen HJG: Notes on the estimation of the numerical density of arbitrary profiles: the edge effect. J Microsc 111:219–223, 1977.

12. Gundersen HJG: Estimation of tubule or cylinder L_V, S_V and V_V on thick sections. J Microsc 177:333–345, 1979.

13. Rasch R: Prevention of diabetic glomerulopathy in streptozotocin diabetic rats by insulin treatment. Glomerular basement membrane thickness. Diabetologia 16:319–324, 1979.

14. Hirose K, Østerby R, Nozawa M, Gundersen HJG: Development of glomerular lesions in experimental long-term diabetes in the rat. Kidney Int 21:689–695, 1982.

part 5

number measurements

This part brings together two chapters describing different methods for estimating numbers of particles from numerical densities (N_V). In Chapter 11, by Reitan, morphometric estimates of bladder cell number are used to assess the value of disaggregation techniques in terms of the yields they offer. In Chapter 12, by de Groot, numbers are an expression of neuronal communication and connectivity mediated by synapses.

Reitan's approach is suitable for the special case where realistic estimates of particle size and shape can be obtained. Other approaches are of general applicability and require no assumptions about particle size or shape. De Groot's method requires some partial serial sectioning, and she explains how the labor involved can be reduced. The method applies where random and independent sectioning cannot solve the problem of estimating particle number.

It should be mentioned too that a new method with great potential has been developed, the "disector" method, which is briefly described in Chapter 1.

cell and nucleus number in intact mouse urinary bladder: a stereological study

Jon B. Reitan

INTRODUCTION

Most cell kinetic studies deal with alterations of the percentages of different cell types in a cell population—cells that accumulate tritiated thymidine, mitotic cells, diploid cells, etc. The total number of cells in a particular tissue is seldom reported.

A valuable tool in cell kinetic studies is flow cytometry.[1] This method analyzes cell type distributions by use of suspensions of single cells or nuclei obtained by various disaggregation methods. The reliability of this fast method is dependent on the cell suspension being representative of the cells in the intact tissue, which is most likely if the yield of cells after disaggregation is high. Farsund[2] worked out a separation method for mouse bladder urothelial cells. The yield was about 10^5 cells per bladder, and he assumed that this corresponded to the total cell number of the bladder. Later, however, Wang et al.[3] obtained 2.5×10^5 cells per bladder and Reitan[4] about 5×10^5 or sometimes up to 10^6 cells per bladder.

From this wide range of total yields it follows that cell type distributions are also uncertain and may lead to false conclusions if the yield of cells related to the actual cell number in the bladders is not known.

The purpose of the present chapter is to study by stereological means the number of nuclei in the mouse bladder urothelium. Since this is the most

The author is greatly indebted to the teachers of the 7th Scandinavian Course in Morphometry and Stereology at Sundvollen, Norway, 1983. The study was supported by the Norwegian Cancer Society and the National Institute of Radiation Hygiene. The expert technical assistance of Mrs. R. Puntervold and Mrs. K. Feren and the typing of the manuscript by Mrs. A. Mortensen are gratefully acknowledged.

exact method of assessing the nucleus number in the intact organ, the values reported can be used to assess the cell yield by various separation methods.

MATERIAL AND METHODS

Female hairless (*hr/hr*) mice (Gamle Bomholt gård, Århus) 10–12 weeks old were used. They were kept eight to a cage under standard conditions with free access to pelleted food (Felleskjøpet, Oslo) and tap water and exposed to artificial light from 0700 to 1900 h. The animals were killed by cervical dislocation. The abdomen was opened immediately, and the bladders were catheterized with a galactography catheter (Danex, Denmark) under visual inspection. The bladders were inflated by hand with 3% glutaraldehyde and 0.1 M sucrose in 0.1 M cacodylate buffer at pH 7.4 to approximately the same size. The bladder neck was ligated and the bladder excised and immersed in the same fixative. After fixation for 2 days, the bladder neck was trimmed off, the fixative content gently expressed and sucked away with filter paper, and the emptied bladder weighed. The emptying was ensured by use of a stereomicroscope to avoid bias due to excess weight. After transfer to cacodylate buffer, the bladders were sliced with parallel razor blades, as indicated in Fig. 1. The rings were then cut into squares of approximately 1 mm^2, successively put into a plastic tissue culture multiwell (Nunc, Roskilde) containing phosphate buffer, and numbered consequently from the bladder dome to the bottom. A systematic sampling procedure starting with a random number was performed, and about 10 blocks from each bladder were selected.

The blocks were rinsed in buffer for 12 h, postfixed in 1% osmium tetroxide, dehydrated in alcohol, and embedded in Epon. Five of the ten blocks from each animal were randomly selected for morphometry. They were cut into 1-μm sections and stained with toluidine blue.

One section from each block was photographed at $40\times$ for an overview and $160\times$ for estimation of nuclei. The slides were projected on a MOP AM 03 digitizing tablet (Kontron, Munich) at final magnifications of 329 and 1317, respectively. At the lower magnification, the urothelial area as a percentage of total tissue area was registered. In some cases the urothelium together with its basement membrane detached from the muscular layer during embedding, probably in the dehydration step, and in these cases the embedding medium without tissue inside the blocks was subtracted. At the higher magnification the number of nuclear profiles per area (N_A) and their maximal profile diameter (d_{max}) were estimated. Profiles were counted according to the "forbidden line" principle as indicated in Fig. 2. The number of nuclei per unit volume (N_V) was calculated using the DeHoff and Rhines formula with Abercrombie's modification[5–8]:

$$N_V = \frac{N_A}{\overline{D} + t} \tag{1}$$

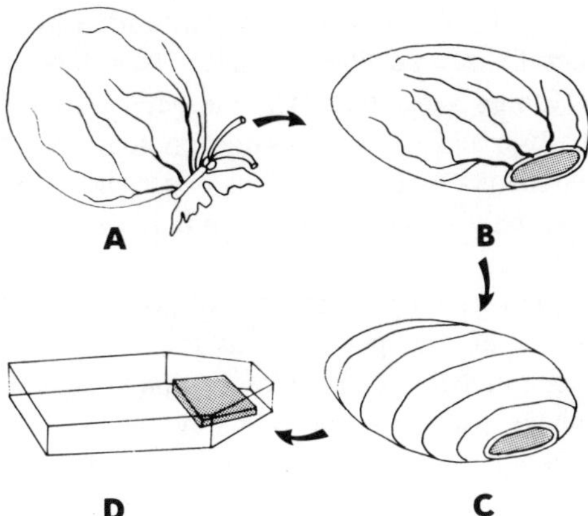

FIGURE 1. Procedure for sampling and embedding. (A) Inflation-fixed bladder ligated at bladder neck. (B) Bladder neck trimmed off. Bladder emptied and weighed. (C) Bladder cut in rings with parallel razor blades. Rings then cut in squares. (D) Sampled square specimen flat-embedded in Epon.

where $\overline{D}$ is the mean caliper diameter and t the section thickness. The mean caliper diameter was determined by subjecting the profile distribution of the urothelial nuclei to a Saltykov correction procedure[8,9] and calculating the mean.

The Saltykov correction procedure, which transforms histograms of profile diameter distribution into histograms of sphere diameter distributions, requires the determination of profile diameters in random sections of spheres. In random sectioning of a specimen containing a number of spheres, the sections always traverse the particles in such a way that the profiles are circles. When oblate ellipsoids are traversed with subsequent section planes parallel to the ellipsoid rotational axis, the d_{max} of the section profiles has the same frequency distribution as if a number of spheres were the same diameters as the true maximal diameters of the ellipsoids were traversed. This is due to the fact that projections of the spheres or ellipsoids result in identical distributions of circles on a plane perpendicular to the axes of the parallel-oriented ellipsoids and to the section plane. Thus, the principles underlying the Saltykov correction procedure are valid in this special situation.

Minor variations from the perpendicular orientation of the section plane to the basement membrane do not seriously affect the determinations of d_{max}. However, as the section plane approaches parallelism to the basement membrane, the section profiles are altered from ellipses to circles. This does not alter the nuclear volume fraction (V_V) but does have some effect on the N_A. However, using the DeHoff and Rhines formula in such a situation would also imply another value of $\overline{D}$, so the N_V would be the same.

Although the inflation fixation was standardized, variations in the tissue stretching do affect the mean caliper diameter. Therefore this value was calculated for each bladder separately. The measurements were performed by two independent observers using the same photographs. Observer A performed the measurements without previous experience; observer B had performed numerous analyses of bladder urothelium but applied the same counting rules as observer A (Fig. 2).

The determination of total nuclear number per bladder is independent of processing shrinkage of the specimens as the reduction in the reference volume or area is counterbalanced by a lower total volume or bladder weight. However, if shrinkage affects the nuclei and cytoplasm to a different degree, or the main shrinkage occurs after the fixation and during embedding, problems may arise. Therefore the quality of the processing regarding shrinkage was checked by measuring the d_{max} of 100 erythrocyte profiles. Erythrocytes are nearly uniform in size and it was assumed that they were sectioned at random. The erythrocyte diameter (D) was calculated using the formula

$$D = \frac{4\bar{d}}{\pi} \tag{2}$$

where $\bar{d}$ is the mean of measured d_{max} values of individual erythrocyte profiles. In these calculations the erythrocytes were analyzed as disks of negligible thickness. The D of the erythrocytes was also evaluated by identifying the breakpoint of a cumulated curve of d_{max}.

RESULTS

The morphometric values obtained are given in Table 1. The total number of urothelial nuclei per mouse bladder was in the range $1-1.5 \times 10^6$. The value obtained by observer A was 70% of that obtained by observer B.

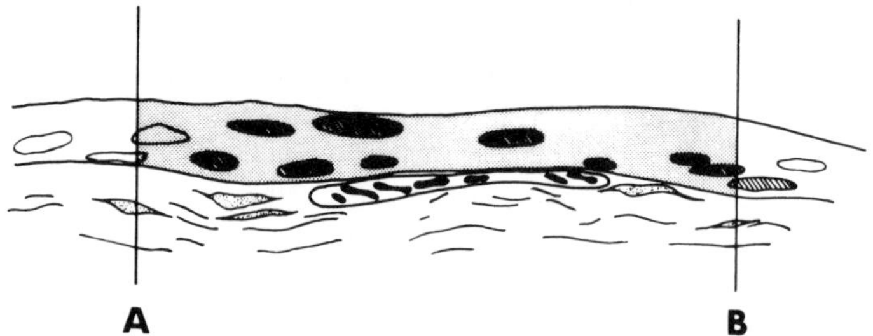

FIGURE 2. Counting principles. Parallel lines A and B with a fixed distance were superimposed randomly on the micrographs. Urothelial reference area is stippled. Nuclear profiles touching line A were excluded, whereas profiles touching line B were included as indicated by hatching.

TABLE 1. Morphometric Values of Normal Mouse Bladders Obtained by Two Independent Observers[a]

| | | Bladder | Urothelium | | Nuclei in urothelium | | | |
No.	Observer	Weight (mg), W_B	Weight (%), $\%_U$	Weight (mg), W_u	Diameter (µm), $\overline{D}$	Profile no., N_A	Nucleus no., N_V	Total no., N_B
1	A	13.6	15.48 ± 1.04	2.07	12.70	6.31 ± 0.48	0.46	0.95
	B		17.80 ± 1.92	2.39	11.76	7.38 ± 0.41	0.58	1.39
2	A	16.4	15.14 ± 1.43	2.43	13.11	5.19 ± 0.50	0.37	0.89
	B		18.00 ± 2.00	2.95	12.40	6.53 ± 1.10	0.49	1.45
3	A	15.1	14.90 ± 1.26	2.25	12.60	5.97 ± 0.66	0.44	0.99
	B		20.75 ± 3.77	3.13	12.10	6.55 ± 0.46	0.51	1.60
4	A	16.5	15.86 ± 3.32	2.62	12.02	6.72 ± 0.70	0.52	1.35
	B		17.25 ± 3.59	2.85	11.56	6.84 ± 0.36	0.54	1.54
Mean	A	15.35 ± 1.45[b]	15.36 ± 0.42[b]	2.34 ± 0.24[b]			0.45 ± 0.06[b]	1.05 ± 0.21[b]
	B		18.45 ± 1.57[b]	2.83 ± 0.32[b]			0.53 ± 0.04[b]	1.50 ± 0.09[b]

[a] W_B, Weight of excised bladder; $\%_U$, percentage of bladder wall consisting of urothelium, mean of 5 blocks ± $SD_{(x)}$; W_U, weight of urothelium of excised bladder, product of W_B and $\%_U$; $\overline{D}$, mean caliper diameter of nuclei; N_A, number of nuclear profiles per area of urothelium, 10^{-3} µm^{-2} ± $SD_{(x)}$; N_V, number of nuclei per volume of urothelium, 10^6 mm^{-3}; N_B, number of urothelial nuclei per excised bladder, 10^6. The Saltykov procedure in determining $\overline{D}$ complicates the calculation of SD, which is therefore omitted. This also applies to the derived values of N_V and N_B. The N_A and $\overline{D}$ values are dependent on the stretching of the bladder during fixation, so total means are not given.

[b] ± $SD_{(\overline{x})}$.

Observer B recorded more small profile diameters than observer A, resulting in an approximately 5% smaller mean caliper diameter. The larger discrepancy, however, is due to the interobserver differences in the percent urothelium of the bladder wall and the number of nuclear profiles per area. Here observer A recorded only 83 and 89% of the values of observer B, respectively.

The diameter of the erythrocytes determined from Eq. (2) was 6.2 µm. The cumulated diameter curve with the breakpoint at class 12 (Fig. 3) also demonstrated a diameter of 6.2 µm. No ring profiles were among the erythrocyte profiles measured, which might have occurred in a true random sectioning of the erythrocytes. In this investigation they therefore behaved more or less as disks.

DISCUSSION

The aim of cell type distribution analysis with flow cytometry is to describe the composition of the intact tissue. For reliable results, the cell suspension must therefore be representative. The only way to study the intact tissue quantitatively is by use of morphometry and stereology.

The value obtained for N_V, about 0.50×10^6 mm^{-3}, is approximately twice the value of N_V of liver cells.[10,11] Taking into account that the liver

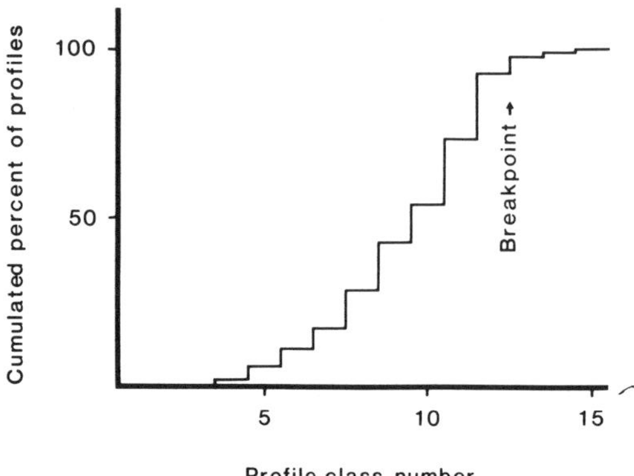

FIGURE 3. Curve of cumulated maximal diameters of erythrocyte profiles sectioned at random. Breakpoint denotes the maximal diameter of erythrocytes in specimen.

cells are extraordinarily large, the value obtained seems to be within reasonable limits.

In this stereologic investigation, the bladder trigone was cut off, thus reducing slightly the total cell number available. However, this is also the case in the isolation methods described by Farsund[2] and Reitan[4] and apparently in the experiments of Wang et al.[3] Thus the cell numbers obtained by these different methods should refer to the same amount of urothelium. The cell numbers obtained by Farsund[2] and Reitan[4] refer to the same mouse strain as that used in this investigation.

The relevance of using the DeHoff and Rhines formula and the Saltykov correction requires some comment. The urothelium is a highly structured tissue with superficial large "umbrella" cells, probably with octoploid nuclei, and deeper cells with diploid and tetraploid DNA content.[12,13] In the unexpanded bladder the nuclei look fairly spherical, but the expansion during fixation seems to transform all the nuclei to oblate ellipsoids or disks lying parallel to the basement membrane (Fig. 4). Because of the use of flat embedding the section plane was perpendicular to the basement membrane (Figs. 1 and 2). No orientation regarding rotational axes was attempted during embedding, so the factor of randomness necessary for use of the formulas was taken care of. The DeHoff and Rhines method involves no assumptions about the form of the particles measured. The formula is generalized to random plane sections but is also valid for transections arranged as in

FIGURE 4. Scanning backscatter electron micrograph of inflation-fixed mouse urothelium, showing binucleated hexagonal surface cell and deeper urothelial nuclei. AgNOR stain. (Courtesy of Dr. Franz Thiebaut.)

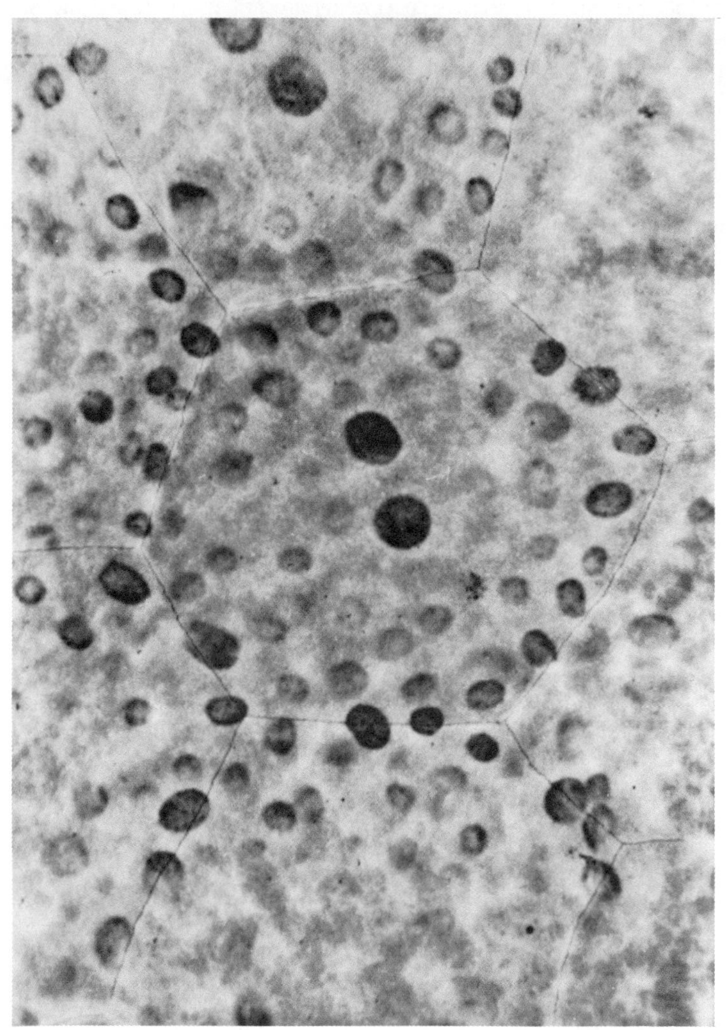

these experiments, provided the $\overline{D}$ is determined in a direction perpendicular to the section plane.[8,14]

In the experiments reported here, the section planes were nearly parallel to the ellipsoid axes (perpendicular to the basement membrane) so the values of the profile diameters need not be corrected. It should be stressed, however, that the nuclei in these experiments are not spheres, and the Saltykov procedure is not generally applicable in N_V determinations. The use of the Saltykov correction procedure has been discussed in detail elsewhere.[8]

Observer A recorded fewer small profiles and profile diameters than observer B. However, the discrepancy in the determination of the percent urothelium of the bladder wall may indicate that some small basal cells and the corresponding area were erroneously excluded by observer A or that areas with small nuclei below the basement membrane were included by observer B.

With toluidine blue stain the basement membrane may be difficult to define. Thus this apparent truncation may be due to suboptimal staining. Later experiments showed that the basement membrane may be more easily defined with other stains. Which of the two sets of data should be regarded as "true" is difficult to assess. The apparent truncation or "lost caps" effect in the data of observer A results in underestimation of N_A and overestimation of $\overline{D}$, in both instances resulting in underestimation of the total cell number. The effect on the Saltykov calculation of $\overline{D}$ is of minor importance compared to the effect on N_A.

A linear shrinkage of some percent should be raised to the power 3 in volume calculations. Thus, the calculations might be altered if there was any marked shrinkage of the specimens during the embedding process and after the weighing that affected the urothelium and bladder muscular wall differently. However, the erythrocyte diameter of 6.2 μm agrees well with the value for the normal mouse erythrocyte at 2 months of age.[15] Moreover, the volume changes during fixation and Epon embedding are reported to be within 3–5%,[6,8] so there is no indication that shrinkage or other processing artifacts influenced the determinations.

In the urothelium, there is obviously a difference between nuclear number and cell number, due to binuclearity. In particular, the surface cells are often binucleated (Fig. 4). With toluidine blue staining the cell margins are hardly visible. Moreover, the surface cells are hexagonal and a Saltykov procedure cannot be used for determination of the mean caliper diameter of the cells. Thus the determinations reported here are not completely relevant for evaluation of cell yield without knowledge of the number of binuclear cells. However, there seems to be reason to believe that the mouse urinary bladder contains about 10^6 urothelial cells, which is more than previously assumed. It can also be concluded that the various techniques for making single-cell suspensions of bladder urothelium do not disperse all the cells in question.

REFERENCES

1. Göhde W, Dittrich W: Impulsfluorometrie—ein neuartiges Durchflussverfahren zur ultraschnellen Mengenbestimmung von Zellinhaltsstoffen. Histochem Ultrastrukt Suppl 10:429–437, 1971.
2. Farsund T: Preparation of bladder mucosa cells for microflow fluorometry. Virchows Arch B Cell Pathol 16:35–42, 1974.
3. Wang CY, Linsmaier-Bednar EM, Garner CD, Lee M: Induction of unscheduled DNA synthesis in primary culture of dog, rat and mouse urothelial cells by arylamine and nitrofuran derivatives. Cancer Res 42:3974–3977, 1982.
4. Reitan JB: Cell numbers and ploidy classes in the normal bladder urothelium in hairless mice. Virchows Arch B Cell Pathol. 48:289–297, 1985.
5. Abercrombie M: Estimation of nuclear population from microtomic sections. Anat Rec 94:239–247, 1946.
6. Aherne WA, Dunnill MS: Morphometry. London: Edward Arnold, 1982.
7. DeHoff RT, Rhines FN: Determination of the number of particles per unit volume from measurements made on random plane sections: the general cylinder and the ellipsoid. Trans AIME 221:975–982, 1961.
8. Weibel ER: Stereological Methods, volume 1. London: Academic Press. 1980.
9. Saltykov SA: Stereometric Metallography, 2d ed. Moscow: State Publishing House for Metals and Sciences, 1958.
10. David H: Quantitative Ultrastructural Data of Animal and Human Cells. Leipzig: VEB Georg Thieme, 1977.
11. Reith A, Barnard T, Rohr H-P: Stereology of cellular reaction patterns. Crit Rev Toxicol 4:219–269, 1976.
12. Hicks M: The mammalian urinary bladder: an accommodating organ. Biol Rev 50:215–246, 1975.
13. Walker BE: Polyploidy and differentiation in the transitional epithelium of mouse urinary bladder. Chromosoma 9:105–118, 1958.
14. Cruz-Orive L: On the estimation of particle number. J Microsc 120:15–27, 1980
15. Crispens CG: Handbook on the Laboratory Mouse. Springfield, Ill.: Charles C Thomas, 1975.

chapter 12

improvements of the serial section method in relation to estimation of the numerical density of complex-shaped synapses

Didima M. G. de Groot*

INTRODUCTION

In estimating the numerical density (N_V) of a type of particles in a tissue sample, both the number per unit area (N_A) and the particle size must be taken into account. The particle size is described by the mean projected height, tangent, or caliper diameter $(\overline{H})$ averaged over all orientations. Under ideal conditions, N_V equals the ratio N_A over $\overline{H}$ [Eq. (2.79) in Ref. 1]. Whereas the determination of N_A is a relatively simple task, the estimation of $\overline{H}$ may cause considerable problems. Most conventional methods for estimating N_V start by making strict assumptions regarding particle shape. Then N_V is estimated on the basis of N_A and a simplified measure of particle size. However, such assumptions may cause serious errors. Cruz-Orive (Ref. 2 and Appendix) described a method for estimating N_V without making any assumptions about size, shape, and orientation of the particles. In essence, this method makes use of serial sections for the estimation of the "effective projected height" $(\tilde{H})$. The quantity $\tilde{H} + T$ is obtained by multiplying the total number of sections in which a particle appears by the thickness of a section.

In studies on synapses in the cerebellum[3] and the hippocampus[4] of the rat, it has been shown that calculations of N_V based on the simple shape assumption that synapses are disklike structures may lead to erroneous biased results. In addition, it appeared that the method of Cruz-Orive must be applied when

* With an appendix by Luis M. Cruz-Orive.

Mr. E. Bierman is gratefully acknowledged for technical assistance and Mr. M. Boermans for photography. Gratitude is expressed to Dr. G. Vrensen (Netherlands Ophthalmic Research Institute, Amsterdam) for constructive comments on the manuscript. Dr. J. Bleichrodt and Dr. O. Wolthuis are acknowledged for critical reading of the manuscript and Mrs. R. Engelen and W. Roelofs for careful typing. This work was supported by Shell Internationale Research Maatschappij B.V.

the numerical density of complex-shaped synapses has to be estimated. A disadvantage of the Cruz-Orive method is that serial sectioning and inspection are rather laborious and time-consuming. The aim of the present chapter is to show how the efficiency of the method can be improved to make it more suitable for routine application. A common problem in serial sectioning is that the size of the pyramid-shaped section is kept very small in view of the necessarily large number of sections within one series.[4] With such small sections the sample size is extremely small, so several series must be cut for quantitative studies. Nevertheless, the statistical relevance of the results obtained from such serial section samples may sometimes be questioned. The use of large sections overcomes this problem to a large extent and considerably reduces the time spent on ultramicrotomy. In the present chapter a serial sectioning technique is described in which large sections are cut.

Another problem in serial sectioning of small tissue samples is the orientation within the tissue, especially when layered or polarized tissues are to be investigated. Large sections, preinspection at the light microscopic (LM) level, the use of specific grids, and the use of a rotation holder and calibrated micrometer with the electron microscope greatly overcome the problem of orientation within the tissue and facilitate recognition of the areas of interest in sequential sections. These points will be dealt with in detail. In addition, the application of the method of Cruz-Orive (Refs. 2 and 5 and Appendix) is shown in a simplified example. The improved method will be illustrated using an example from the author's laboratory on the synaptic density in the hippocampus of the rat as calculated with Cruz-Orive's formulas (Appendix). The advantages and disadvantages inherent in the application of the whole procedure are discussed. To avoid an excess number of illustrations, the hippocampus is also used to illustrate the general aspects of the serial sectioning and sampling technique.

GENERAL PROCEDURES

Fixation, Tissue Sampling, and Embedding

After a proper perfusion fixation, e.g., according to Peters,[6] the brain is taken from the skull. Brain structures are dissected (Fig. 1a) and stored in buffer. The structures in one hemisphere are used for quantitative studies. In some instances it can be important to use the contralateral counterpart to measure the reference volume of the particular brain structure. Slabs of tissue are removed from the brain structure by making transverse cuts of approximately 4 mm with a set of razor blades (Fig. 1b). Depending on the size of the brain structure to be investigated, three to five slabs containing all the layers are selected randomly for further quantitative studies. From the appropriate tissue samples, 75-μm slices are cut on a Vibratome (Oxford Instruments) (Fig. 1c). The Vibratome slices can be examined under a dissecting microscope when it is desirable to sample distinct areas within the slice (Fig.

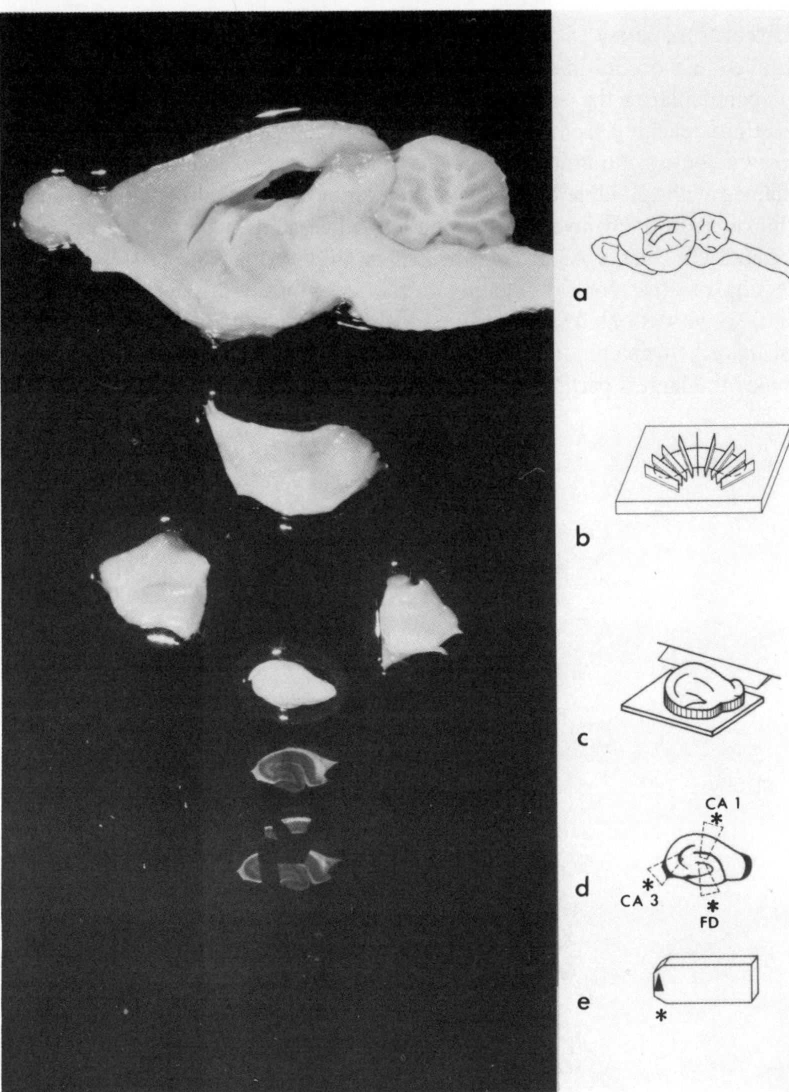

FIGURE 1. Tissue sampling, illustrated for the hippocampus. Embedding of the tissue samples occurs in an orientated way so that cutting with the ultramicrotome starts perpendicular to the outer surface—i.e., the ventricular surface for the hippocampus—of the brain structure (note asterisks). See text for further details. FD, Fascia dentata.

1d). After thorough rinsing in buffer, the slices can be block stained. Embedding is carried out in an orientated way in flat embedding molds (Fig. 1e).

Ultramicrotomy

Embedding occurs in such a way that ultramicrotome cutting is carried out perpendicular to the outer surface of the brain structure. Semithin (500-nm) sections reaching from the surface down to the center of the brain structure are cut on an ultramicrotome using glass knives. These sections are post-stained with toluidine blue. They are examined in a light microscope and the thickness of each layer is measured as illustrated in Fig. 2 for the hippocampal CA1 and CA3 areas. These data are essential for the subsequent selection of areas for examination in the electron microscope. Ultrathin (50-nm) serial sections from the same blocks are cut using a diamond knife. The distance between the first and the last sections of a series should be more than twice the largest particle diameter (Ref. 2 and Appendix, note N2). The sec-

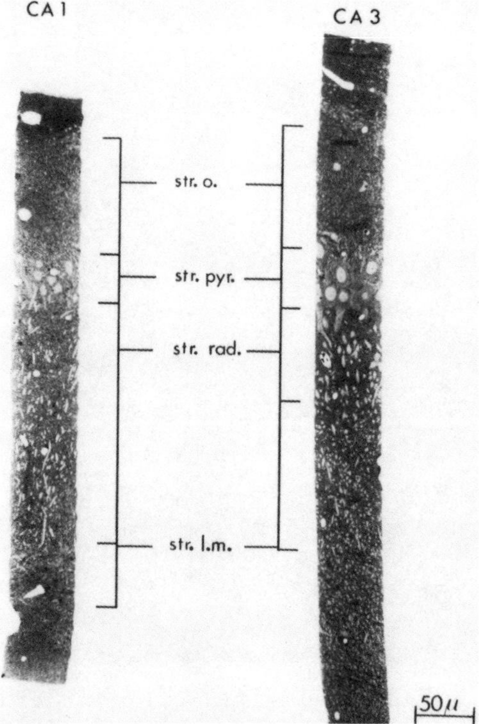

FIGURE 2. Semithin toluidine blue-stained sections of the hippocampus, illustrating the thickness of the distinct layers within the CA1 and CA3 areas. Str. o., stratum oriens; str. pyr., stratum pyramidale; str. rad., stratum radiatum; str. l. m., stratum lacunosum moleculare.

tions are 75 μm wide—the thickness of the Vibratome slice—and can be up to 1500–2000 μm high. Several methods are available for measuring the thickness of the sections, e.g., the Small fold method,[7] densitometry and interferometry,[8–10] and comparison of exposure densities with those of calibrated specimens in the electron microscope.[11–13]

In the Small fold technique the first and last three to five sections adjacent to the series collected for quantitative analysis of the particles of interest are used. The other techniques can be applied to all sections of the series. Serial sectioning is carried out as shown in Fig. 3. The letters of the list below correspond to the letters in the figure.

(a) The specimen block is trimmed in such a way that the bottom edge of the block is orientated parallel to the knife edge during sectioning. However, in contrast to what is usually done, the top edge of the block is not parallel to the bottom edge but is arrow-shaped (Fig. 4). As a result, the sections cut are in contact with the knife edge at only a single point. Consequently, they can easily and quickly be removed one by one from the water surface in the knife boat.

(b) The time interval between two subsequent sections is kept as long as possible by using the lowest speed of the reverse stroke of the ultramicrotome. This allows just enough time to pick up each section individually from the water surface with the aid of a platinum loop held between a pair of tweezers.

(c) The platinum loop containing the section on a water surface is carefully placed on Pyoloform-FN50 (Wacker Chemie) coated and systematically coded glass slides. To avoid damaging the support film, the loop is "dropped" when it is a few millimeters above the support film.

(d) The water in the loop is drawn off by touching the outer edge of the loop with a wedge-shaped piece of filter paper, thereby depositing the floating section on the Pyoloform film. The section is then alowed to dry. Subsequently, the loops are dropped off the slides by turning the slides upside down. If necessary, the sections can be poststained in a tightly closed petri dish. Drops of stain, e.g., lead or uranyl, are then placed over the individual sections and are flushed off with distilled water after a proper staining time.

(e) The Pyoloform support film carrying the sections is removed from the slide by stripping off onto a water surface.

(f) Grids are then placed over the sections under a dissecting microscope in an orientated way while the support film is still floating on the water surface. Slotted grids (pitch 250 μm, hole 200 μm, and bar 50 μm; VECO, R100A, Cu 3.05 mm; 175, 841) from which one bar has been dissected under the dissecting microscope are used.* As shown in Fig. 5,

* Currently, these grids are produced commercially by Oscar Stolk Scientific, Capelle a/d ijssel, The Netherlands.

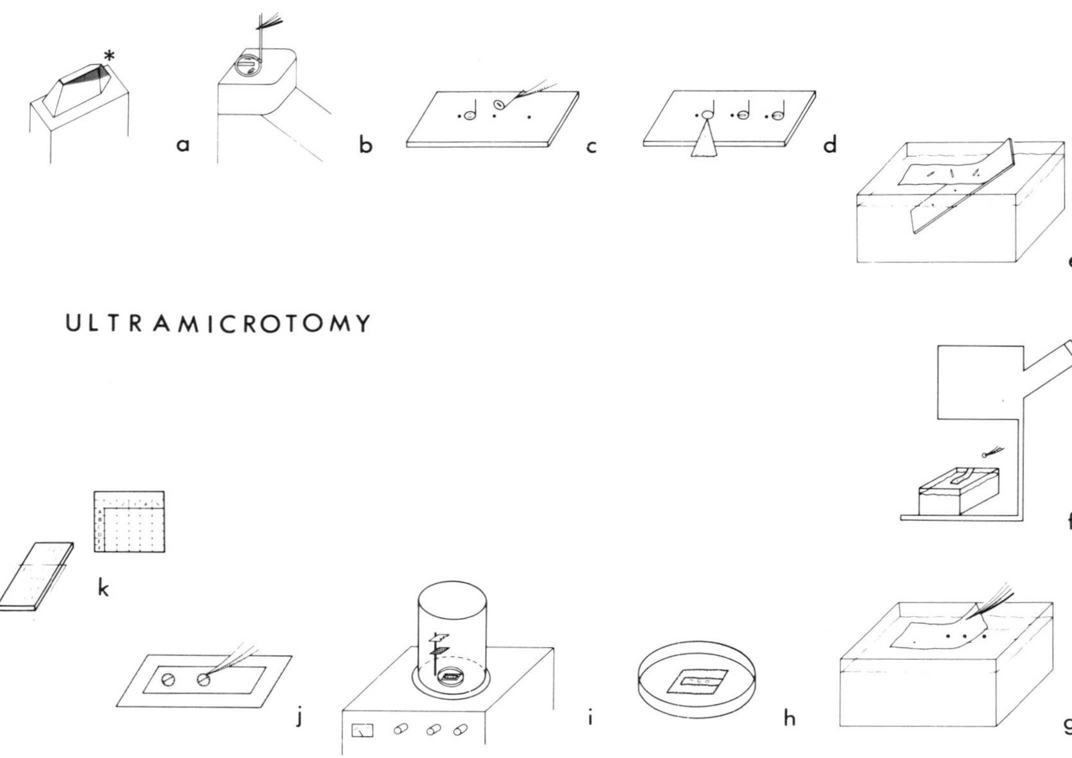

FIGURE 3. Serial sectioning procedure. For details see the section on "Ultramicrotomy." (a) *, Outer surface of the brain structure.

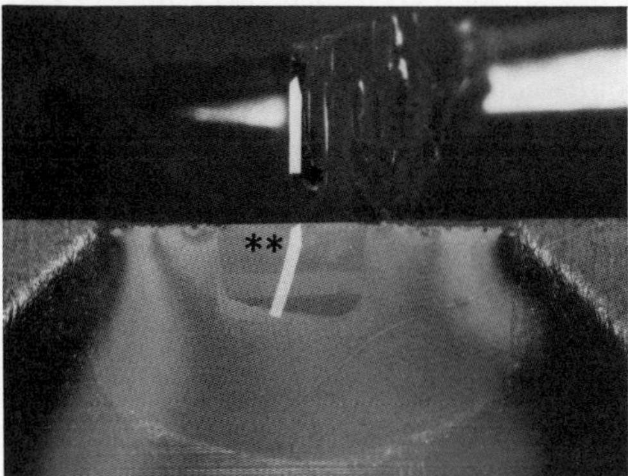

FIGURE 4. Survey picture of a specimen block mounted in the ultramicrotome, showing that the top edge of the pyramid-shaped specimen block is arrow-shaped. Note that the bottom edge of the block (∗) is parallel to the knife edge and that the cut section (∗∗) is in contact with the knife edge at a single point only.

the oblong section is placed with its longitudinal axis between and parallel to the two bars adjacent to the removed one, i.e., on the place of the removed bar.

(g) The Pyoloform support film with the overlying grids is picked up from

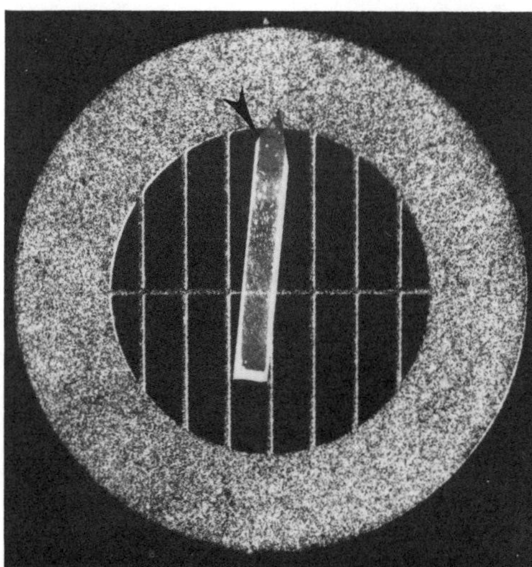

FIGURE 5. Specimen grid, illustrating the carefully oriented position of the section with respect to the grid bars. (Arrowhead indicates the position of the removed bar.)

the water surface with a piece of Parafilm ("M," American Can Company, Greenwich, Conn.).

(h) The piece of Parafilm carrying the Pyoloform film with the sections and grids is put in a closed petri dish with the Parafilm side down.

(i) After thorough drying over silica gel, the grids are vacuum-coated with a thin layer of carbon.

(j) The grids are detached from the Parafilm with a pair of tweezers. It is important to encircle the grids with the tweezers before detaching them. By doing this, the part of the support film covering the grid is separated from the rest of the film. Hence, the possibility that the film with the section will be detached from the grid is avoided.

(k) The grids are stored in a grid box and are numbered.

Electron Microscopic Sampling

The ultrathin sections are inspected in a conventional transmission electron microscope equipped with a goniometer attachment and a rotation specimen holder. The general procedure for selecting and photographing areas of interest is carried out as shown in Fig. 6. The letters of the list below correspond to the letters in the figure.

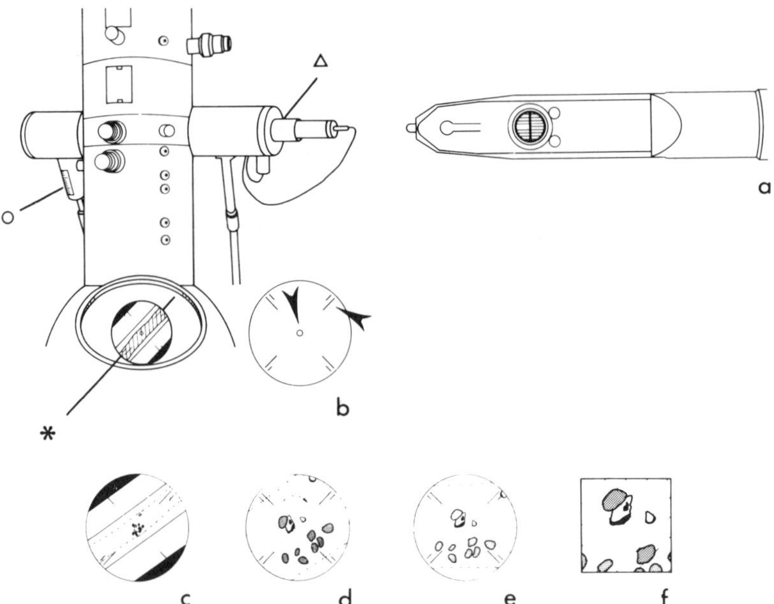

FIGURE 6. Strategy for electron microscopic sampling of serial sections. (a) Note standardized position of the grid with the removed grid bar. (b) *, Direction of movement of micrometer. Note that the longitudinal axis of the oblong section is in the axis of the direction of movement of the micrometer. ○, Calibrated micrometer; △, goniometer attachment and rotation holder; ►, screen markers and center point on the main screen. See text for further details.

(a) The grid is put in the specimen holder. A standardized position greatly facilitates the orientation in the sequential sections.

(b) At a scanning magnification (e.g., 200× on the main screen) the grid is rotated to a position in which the longitudinal axis of the section is in the axis of the specimen traverse control (Fig. 6b, *) to which the micrometer is connected (Fig. 6b, ○).

(c) The procedure is started in a layer that can easily be recognized at low magnification by the presence of specific landmarks, e.g., cell nuclei. The center of this layer is moved to the center of the main screen. The micrometer value (Fig. 6b, ○) is read and noted. This "starting position" is subsequently used as reference to find other areas of interest that cannot be recognized as such at low magnification.

(d) The magnification is increased to a level appropriate for counting and measuring the structural elements of the particles of interest.

(e) While keeping the center point of this area in the center of the screen, the grid is rotated to another position, i.e., the position in which the upper and lower borders of the layer run parallel to the upper and lower edges of the photograph to be taken. The resulting image on the main screen is drawn diagrammatically. Special attention is paid to the relative position of larger structures, such as blood vessels, cell nuclei, and myelin sheets, with respect to the screen markers (Fig. 6b, arrowheads). Large structures, of course, reappear in a large number of sections and will facilitate orientation in the sequential sections of the series.

(f) The image is photographed. It appears to be helpful for the final analysis to list in a simple table whether or not a particular area was photographed and not missed due to, e.g., folds or dirt on the sections.

Knowing the thickness of each layer from the light microscopic measurements (cf. Fig. 2), any area within a certain layer can be selected. If desired, steps b and c are then repeated in order to return to the starting position. From this starting position the image is moved to reach another area of interest. The distance to be covered is known from the light microscopic measurements. This distance is read on the micrometer, which is connected to the specimen traverse control. Subsequently, steps d to f are repeated. After sufficient areas within one section in different layers have been photographed, an adjacent section is taken and put in the specimen holder in the standardized position and the procedure is repeated. The drawings obtained in step e are used to select and to photograph identical areas in the sequential sections of the series.

The same micrographs can be used for quantitative analysis of much smaller objects, which, although they can be counted, need much higher magnification to be measured. To achieve this, the micrographs are placed on an illuminated screen and a video camera is placed over them. The enlarged image is projected on a monitor. To facilitate orientation within the micrograph after enlargement on the monitor, a homemade square lattice was found to be very useful. The image on the monitor represents one of the numbered square areas of the lattice (Fig. 7).

Stereological Analyses

The parameter N_V is estimated according to the method of Cruz-Orive (Appendix).

To illustrate the application of this procedure and the formulas, a simplified imaginary example on more or less "curved" synapses of different sizes is shown in Fig. 8. A series of nine sections is drawn with a reference area (A_i) of 225 μm^2. All synaptic contact zones (N_i) in the center section—i.e., the fifth section—are counted, applying Gundersen's unbiased counting rule.[14] This results in six synaptic contact zones. Two of these synaptic profiles are serially tracked. The quantity $\tilde{H}_{ij} + T$, where $\tilde{H}_{ij}$ is a particle's "effective projected height" (see Appendix) is estimated by multiplying the number of sections in which each particle appears (m_{ij}) (six and four, respec-

FIGURE 7. Equipment for enlargement of low-power micrographs connected to a semiautomatic image analysis system (Videoplan, Kontron Messgeräte, Eching, FRG). The micrographs on the illuminated screen are covered with an appropriate square lattice. One square is magnified with a video camera and the image is projected on the screen of the monitor. The black lines on the screen of the monitor correspond to the edges of the square of the lattice; the white lines delineate the "active area" of the Videoplan. *, Enlarged transversely cut synaptic contact zone after E-PTA treatment.

tively) by the mean section thickness T_i. The section thickness (T_i) is measured by the Small fold technique[7] and has a mean value of 42 nm.

Data from three other series of the same imaginary brain structure are given, making a total of $m = 4$ series. Values of $estN_V(\text{syn, struc})$, $SE\{est(N_V)\}$, and $CE\{est(N_V)\}$ are calculated using the formulas mentioned in the Appendix. The detailed calculations are shown in Fig. 8.

APPLICATION

A 30-month-old rat was fixed by transcardial perfusion with glutaraldehyde/paraformaldehyde mixtures according to Peters.[6] Both hippocampi were dissected and stored in sodium cacodylate buffer.[6] The right hippocampus was used to estimate the reference volume; the left hippocampus was used for the quantitative study of synapses. Slabs of approximately 4 mm were removed from the left hippocampus by making transverse cuts with a fan-shaped set of razor blades (cf. Fig. 1). This resulted in five slabs of the banana-shaped hippocampus. The two outer slabs did not contain all hippocampal layers. Two other slabs, each containing all the hippocampal layers, were taken for the quantitative studies. Vibratome slices of 75 μm were examined under the dissecting microscope and the CA3 area was dissected for investigation. After thorough rinsing in sodium cacodylate buffer, the slices were block-stained with ethanolic phosphotungstic acid (E-PTA) according to Vrensen and De Groot.[15] With this method, synaptic contact zones are selectively stained, which facilitates their recognition at the electron microscopic level. Semithin and ultrathin serial sections were cut with a Reichert Om U3 ultra-microtome, perpendicular to the ventricular surface and reached to at least the stratum lacunosum moleculare. The semithin toluidine blue sections were examined under a light microscope and the thicknesses of the four distinct layers, i.e., stratum oriens, stratum pyramidale, stratum radiatum, and stratum lacunosum moleculare, were measured with the Videoplan (Kontron Messgeräte, Eching, FRG; Fig. 2). Serial sectioning and photography were carried out as described in the general procedures.

In this study on hippocampal synapses, it was established that the largest synapse had a diameter of approximately 600 nm. Assuming a section thickness of 50 nm on the average, series of 30 sections were collected. The section thickness was estimated by the fold method proposed by Small.[7] For these measurements the first and last three to five sections adjacent to the series of 30 were collected on Pyoloform carbon-coated grids instead of glass slides. The grids were pressed on sections when floating on the water surface in the knife boat. This manipulation caused folds in the section. Small folds in the section that had a parallel outline over a certain distance were photographed at a final magnification of 56,000×. If a fold stands upright, its width equals twice the section thickness and can easily be measured. The ultrathin sections were viewed in a Philips 201 transmission electron micro-

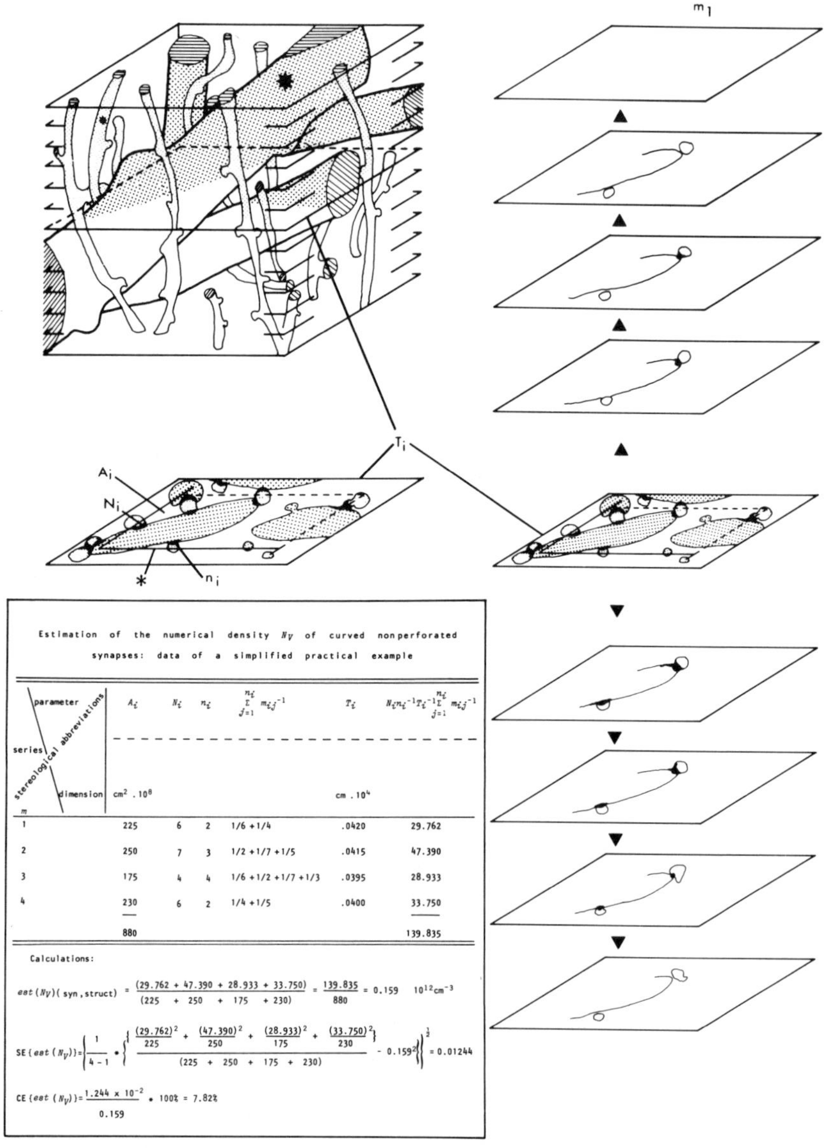

series \ stereological abbreviations \ parameter	A_i	N_i	n_i	$\sum\limits_{j=1}^{n_i} m_{ij}^{-1}$	T_i	$N_i n_i^{-1} T_i^{-1} \sum\limits_{j=1}^{n_i} m_{ij}^{-1}$
dimension	$cm^2 . 10^8$				$cm . 10^4$	
1	225	6	2	1/6 +1/4	.0420	29.762
2	250	7	3	1/2 +1/7 +1/5	.0415	47.390
3	175	4	4	1/6 +1/2 +1/7 +1/3	.0395	28.933
4	230	6	2	1/4 +1/5	.0400	33.750
	880					139.835

Calculations:

$$est(N_V)(\text{syn,struct}) = \frac{(29.762 + 47.390 + 28.933 + 33.750)}{(225 + 250 + 175 + 230)} = \frac{139.835}{880} = 0.159 \quad 10^{12} cm^{-3}$$

$$SE\{est(N_V)\} = \left\{ \frac{1}{4-1} \cdot \left\{ \frac{\frac{(29.762)^2}{225} + \frac{(47.390)^2}{250} + \frac{(28.933)^2}{175} + \frac{(33.750)^2}{230}}{(225 + 250 + 175 + 230)} - 0.159^2 \right\} \right\}^{\frac{1}{2}} = 0.01244$$

$$CE\{est(N_V)\} = \frac{1.244 \times 10^{-2}}{0.159} \cdot 100\% = 7.82\%$$

FIGURE 8. Simplified example of application of the method of Cruz-Orive (Ref. 2 and Appendix) for estimation of the numerical density (N_V) of particles in a structure without making assumptions regarding the size, shape, and orientation of the particles. A block of brain tissue containing large (✲) and small (∗) neurites is shown. Exchange of information between these nerve cell elements occurs via synaptic transmission. The shape of the synapses to be studied does not resemble the ideal shape of a flat circular disk. Their surface is more or less "curved" although not "perforated." To study the numerical density of these synapses, parallel serial sections are cut from the tissue block. In the center section T_i of a series—

scope equipped with a goniometer and a rotation holder. The starting position in the electron microscopic examination (step c) was chosen in the center of the pyramidal layer (Fig. 6). This layer can easily be identified at low magnification by the presence of the large nuclei of the pyramidal cells. The magnification used in step d was approximately 6000× on the main screen. The image in step f was photographed on 35-mm film (Kodak FGP). The electron micrographs were printed at a magnification of 11×, resulting in a final magnification on the micrograph of approximately 6000×. The exact magnification for each series of micrographs was calibrated using a grating replica (2160 lines/mm). The specimen area included in a micrograph was approximately 50×50 μm^2. In the example presented in this chapter, per slab one micrograph was taken per layer, i.e., stratum oriens, stratum pyramidale, stratum radiatum, and stratum lacunosum moleculare. Stratified systematic sampling was carried out within one micrograph. As will be outlined below, one to five areas per micrograph were selected for the analyses of the synapses. Both "classical" disklike[16] and larger complex-shaped "perforated" synapses were found in all four layers of the CA3 area. These perforated synapses[17] have a presynaptic grid of regularly arranged dense projections, which is partly disrupted.[4] Since these complex-shaped synapses are thought to have a particular functional significance,[17,18] it was found desirable to distinguish between the disklike nonperforated and the complex-shaped perforated synapses. However, only a small portion (5–30%) of the total synapse population of the hippocampus has been shown to be perforated. By examination of only the center section of a series of micrographs for the analysis of the synapses, it cannot be established whether the profile observed belongs to a perforated or a nonperforated synapse. Therefore, all synaptic profiles present in the center section were tracked through the series to obtain a valid estimation of $\tilde{H}$ of the perforated as well as the nonperforated synapses (e.g., see Fig. 9). Stratified systematic sampling was carried out within each micrograph. To be able to do this, a predetermined regular lattice was placed over the micrograph (Fig. 10). The analysis of the synapses was always started in area 3, situated in the center of the micrograph. Subsequently, areas 1 and 5, situated closer to the upper and lower border of the layer, were analyzed. Then if fewer than a dozen synapses—nonperforated as well as perforated—were analyzed the analysis was continued in areas 2 and 4.

---

in the example the series (m_1) is taken approximately at the middle of the tissue block—all synaptic contact zones N_i are counted using Gundersen's[14] frame (∗) for unbiased counting. The straight lines of this frame indicate Gundersen's "forbidden line"; synapses lying on the edge of reference area A_i are counted, provided they do not touch the forbidden line. Some of the synapses (n_i) within A_i are tracked forward (▼) and backward (▲) in the adjacent sections of the series to obtain an estimate of the effective projected height of the synapses. For simplicity, only synapses that are tracked (n_i) are drawn in the adjacent sections of the series. For further details, see the Appendix and the section on "Stereological Analyses."

Three to five areas per layer within one micrograph were thus analyzed, except for the stratum pyramidale, which contains large cell bodies. In this case it was found necessary to analyze in one slab the total area of the micrograph. After obtaining a reliable approximation of $\tilde{H}$ for both types of synapses, N_V was calculated. Since all synaptic profiles in the center sections were serially traced, n_i (number of synapses tracked) equals N_i (total number of synapses present in the reference area). Hence, the formula used for the estimation of $N_V(\text{syn})$ was (see Appendix)

$$\text{est}(N_V) = \frac{\sum_{i=1}^{m} T_i^{-1} \sum_{j=1}^{n_i} M_{ij}^{-1}}{\sum_{i=1}^{m} A_i}$$

where

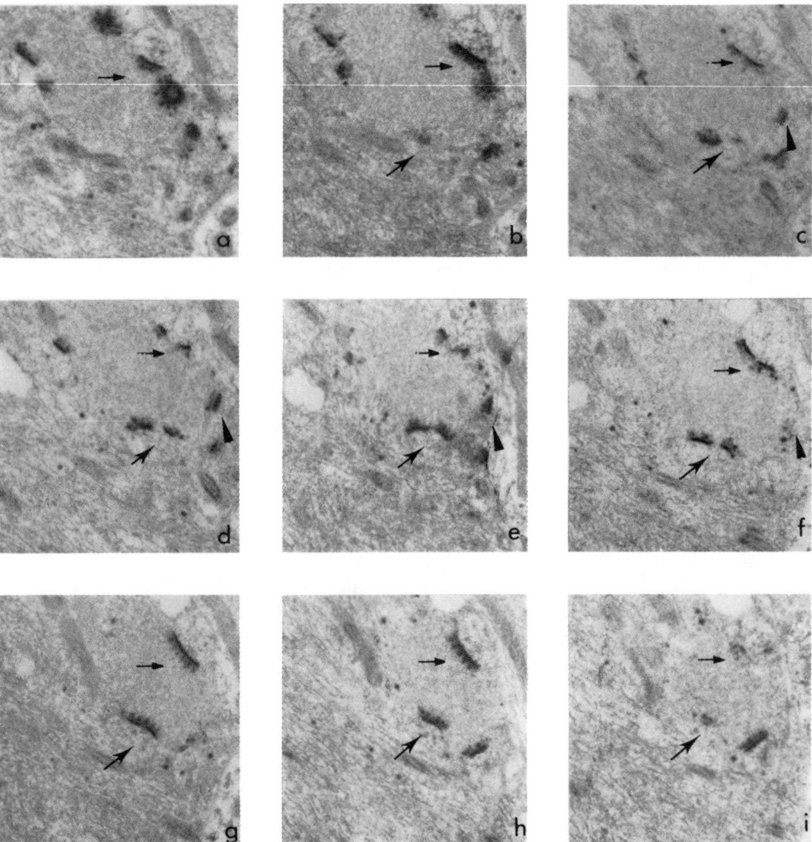

FIGURE 9. Example of E-PTA-treated serial sections in the stratum radiatum of the hippocampal CA3 region of the rat, showing two transversally cut perforated (arrows) and a nonperforated (arrowheads) synapse.

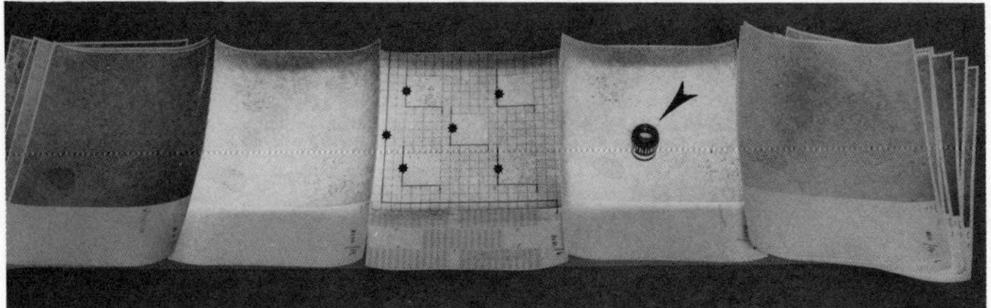

FIGURE 10. Analysis of a series of micrographs for estimation of the numerical density (N_V) according to Cruz-Orive (Ref. 2 and Appendix). Five square areas (1–5) were analyzed using Gundersen's unbiased counting rule (∗, Gundersen's frame). Putting the micrographs on an illuminated screen greatly facilitates the measurements. Areas in the sequential sections corresponding to the five analyzed reference areas are identified by putting the adjacent micrograph on top of the previous one. In this way exactly identical areas are analyzed in the sequential sections. A loupe (arrowhead) may be helpful; for example, in the case of the hippocampal synapses it is used to distinguish, in case of doubt, between perforated and nonperforated synapses or to count the number of presynaptic dense projections along the corresponding synaptic contact zone.

- N_V = number of particles per unit volume
- m = number of series
- n_i = number of particles counted within A_i and tracked through the series
- T_i = thickness of the sections in the ith series
- m_{ij} = number of sections in which the jth tracked particle appears in the ith series
- A_i = reference area

The standard error of est(N_V) is estimated as follows [Appendix, Eq. (A.2)]. Setting

$$y_i = T_i^{-1} \sum_{j=1}^{n_i} m_{ij}^{-1}, \qquad x_i = A_i$$

gives

$$\mathrm{SE}\{\mathrm{est}(N_V)\} = \left\{ (m-1)^{-1} \left[\frac{\sum_{i=1}^{m} y_i^2 x_i^{-1}}{\sum_{i=1}^{m} x_i} - \mathrm{est}(N_V)^2 \right] \right\}^{1/2}$$

The coefficient of error CE$\{$est(N_V)$\}$ was estimated as described in the Appendix and general procedures.

Table 1 shows the results of this study, from which it can be concluded that the different layers of the hippocampal CA3 region contain different numbers of perforated and nonperforated synapses. These regional differ-

TABLE 1. Number of Different Types of Synapses per Unit Volume N_V in Distinct Layers of the Hippocampal CA3 Region of a 30-Month-Old Rat, Obtained from Serial Sections[a,b]

		Parameter	All synapses				Nonperforated synapses				Perforated synapses			
Layer	m	$\Sigma_{j=1}^{m}A_j$ (μm^2)	n_i	[Mean]	SE	CE	n_i	[Mean]	SE	CE	n_i	[Mean]	SE	CE
Stratum oriens	6	510.6	196	2.766	0.349	12.6	177	2.620	0.309	11.8	19	0.146	0.045	30.5
Stratum pyramidale	5	2875.4	252	0.650	0.132	20.4	231	0.623	0.126	20.2	21	0.027	0.007	26.9
Stratum radiatum	5	425.5	171	3.517	0.426	12.1	136	3.148	0.434	13.8	35	0.368	0.073	19.7
Stratum lacunosum moleculare	8	716.1	206	1.718	0.374	21.8	190	1.647	0.372	22.6	16	0.071	0.025	36.0

[a] m, series; A_j, reference area; n_j, number of particles with A_j tracked through the series; [mean], mean number of synapses per unit volume $N_V(10^{12}$ cm$^{-3})$; SE, standard error of the mean; CE, coefficient of error of the mean. The parameters were calculated using the formulas given in this chapter (Appendix).

[b] The CE[mean] in the present example is constant although quite large due to the regional variations within the hippocampus. These variations are mainly the result of including cell bodies and main neurites in our samples. A discussion of that point is presented elsewhere.[31]

ences in perforated and nonperforated synapses cannot be fully appreciated in random sections, because perforated synapses are only partly recognized as such. This leads to biased estimates of N_V of the total synapse population, as was shown by De Groot and Bierman[4] and will be discussed below.

COMMENTS AND DISCUSSION

Methods requiring ultrathin serial sectioning are laborious, which greatly reduces their use for routine application. However, for special purposes serial sectioning may be essential, for example, to estimate the numerical density of nonconvex particles, as shown for complex-shaped synapses in brain tissue by Verwer and De Groot[3] and De Groot and Bierman.[4]

The present chapter shows how the efficiency of the sampling procedures in such stereological studies can be improved. The method described is especially useful in the study of layered structures, which are common in the central nervous system (CNS). The lack of homogeneity of CNS tissues leads to major sampling problems. Adequate sampling can be carried out only at low magnification, which, in turn, introduces problems with respect to discrimination of structural details. So far, this problem has been solved by the use of composites of high-power micrographs. In the procedure described here, electron micrographs are made at low magnification to include large areas. The necessary enlargements for the study of the particles of interest are obtained by means of a video camera. In the example presented, these particles are the substructures of synaptic contact zones. To compensate for regional differences, the tissue blocks are sampled from the organ by a systematic sampling procedure. Stratified systematic sampling is then applied within each layer, which excludes overlap of fields and thus prevents double sampling. The distance between the two borders of a layer is measured in advance at the light microscopic level to make sure that the micrographs are representative for a particular layer. Although this measurement is an extra step in the whole procedure, it is worth the effort since it provides the basis for an unbiased sampling of micrographs at the electron microscopic level. Moreover, this extra step takes little time and may, additionally, lead to the detection of changes in thickness of layers.

Orientation within the section is also improved by including in the section all layers in between the outer surface of the organ and its center. From the histological organization of the neuropil—especially the orientation of large neurites—it can then easily be deduced whether or not the section is cut perpendicular to the outer surface of the organ.

Ultramicrotomy of such large blocks is facilitated by the use of 75-μm Vibratome slices that are embedded in an orientated way and subsequently cut perpendicular to the outer surface of the organ, i.e., the ventricular surface for the hippocampus. Serial sectioning of such large oblong sections is made reliable by trimming the specimen block in such a way that the top edge of the block is not parallel to the knife edge. In contrast to conventional ultramicrotomy, this results in contact of the section with the knife edge at a single point only. The resulting minimal contact, together with a low speed of the reverse stroke of the ultramicrotome and the use of loose platinum loops, guarantees one-by-one removal of a series of sections from the knife boat in the proper sequence. After some practice, complete series of 40 sections or even more can be obtained easily and rather safely.

Orientation in the electron microscope is facilitated by placing each section in an orientated way on a suitable grid. The use of a rotation holder and a calibrated micrometer guarantees quick location of the areas of interest, keeping in mind the measurements of layer thickness. This procedure reduces the time during which the grid is in the electron beam, and illumination and emission can be kept low. Consequently, the damage to sections is minimal and very few are lost. The survival rate of the sections is also enhanced by choosing the best fitting grid with respect to grid bars and holes. On the one hand, the bars should not cover the areas of interest in the section; on the other hand, the holes should not be too large since the support film with the section will then collapse as soon as the grid is brought into the electron beam. Pyoloform support films are much stronger under the electron beam than, e.g., Formvar, collodion, or Parlodion films, but they tend to show more drift. This drift can be prevented by coating with carbon, which has a stabilizing and strengthening effect on the supporting film and the section.

By applying these procedures, a drastic reduction is achieved in the time required for sampling, filing, and documenting the tissues as well as in the cost of photographic materials. These modified procedures bring stereological methods involving serial sectioning within reach of routine application.

With respect to estimation of the numerical density using serial sections, the following critical statements can be made. Conventional derivations of the numerical density of a population of particles (N_V) involve strict assumptions with respect to the shape of the investigated particles. For synapses, it is generally accepted that their shape resembles a flat circular disk.[16] However, deviations from the ideal disk shape, e.g., curved or complex-shaped synapses, have been described in several papers.[3,4,17–25] Synapses may not only be curved or complex-shaped but may also be perforated; i.e., one or more parts of the pre- and postsynaptic surface area may be free of dense projections and postsynaptic densities. This gives the impression that the synapses are perforated.[17,24] It has been suggested by some investigators that these complex-shaped synapses have a particular functional significance since their number has been found to vary. Greenough et al.[17] found an increase in the occipital cortex when rats were reared in an ''enriched'' environment. Vrensen and Nunez-Cardozo[18] found an increase in perforated synapses in the visual cortex of rabbits after visual training. In the present study of rat hippocampus, large complex-shaped synapses were present side by side with smaller disklike synapses, as was found in the visual cortex of the rabbit.[18,24,25] In the latter studies, semithin (500-nm) E-PTA sections were used to reconstruct a frequency distribution of the synapses. Unfortunately, the authors disregarded the fact that they were underestimating the number of the larger complex-shaped synapses. They counted only synapses that were fully incorporated in the sections. Obviously, larger synapses have a greater chance to be cut by sectioning and, therefore, a smaller chance to be fully incorporated in the section; hence they have a smaller chance to be counted. Moreover, Vrensen and Nunez-Cardozo[18] calculated the numerical density of the synapses by the method of Abercrombie[26] based on the disk-shape assumption. As shown by Verwer and De Groot,[3] the validity of this assumption and consequently of such results is questionable. In a study on undernutrition, they compared the results for synaptic density of curved synapses in the cerebellum obtained by conventional methods[26,27]—based on the disklike shape assumption—with those obtained by a method that makes no assumptions regarding size, shape, and orientation of the synapses (Ref. 2 and Appendix). Erroneous results were obtained when methods were applied assuming synapses to be flat disklike structures. The problem is even more complicated when part of the synapses are perforated, as in the study presented in this chapter. It was previously shown by De Groot and Bierman[4] that the simultaneous presence of larger complex-shaped perforated and disklike nonperforated synapses in the hippocampus of rats resulted in underestimation of the synaptic profile length and overestimation of synaptic density when measured in randomly selected ultrathin E-PTA sections. Since the diameter of the presynaptic grid is 100–500 nm, the synapses will mostly be transversely cut in ultrathin (50-nm) sections. In a 50-nm osmicated section, the perforated synapses seem to have an interrupted contact zone when the cut runs through the perforation. Only in that case can it be distinguished from the disklike nonperforated synapse, which lacks such an interruption. In the study by Greenough et al.,[17] carried out with random osmicated sections, only synapses that had the perforation in the plane of the section were considered to be perforated. Synapses without perforation in the plane of the section were erroneously counted as nonperforated synapses.

In E-PTA-treated 50-nm random sections, an interrupted contact zone often cannot be recognized as such and, instead of being counted as a perforated synapse, is counted as two or more separate synapses.[4] The differences in synaptic density and synaptic profile length found between the results obtained from random E-PTA and random OsO_4 sections[15,28,29] can be ascribed, at least in part, to the presence of complex-shaped perforated synapses in the tissue. In addition, the reaction of E-PTA with the paramembranous densities most likely differs from that of OsO_4, resulting in less extensive postsynaptic densities.

To estimate the numerical density in synapse populations with a considerable number of perforated or irregularly shaped synapses, conventional methods are inadequate. Serial sections should be used to distinguish between perforated and nonperforated synapses, and for the calculation of N_V a method should be applied that makes no assumptions about synaptic size and shape. For several reasons, outlined in this chapter, the method of preference is the one introduced by Cruz-Orive (Ref. 2 and Appendix). With this method the effective projected height $\tilde{H}$ of the synapses is estimated directly from serial sections.

As shown in the imaginary example on curved synapses (Fig. 8), only some of the synapses (20 on the average) counted in the center section of the series must be tracked through the series in order to derive a valid estimate of their effective projected height. However, in the example presented in this chapter all synapses counted in the center section had to be tracked in order to distinguish between perforated and nonperforated synapses. A three-dimensional reconstruction is not necessary.

Complex-shaped particles are not overestimated since the particles are tracked through the series of sections. Since no assumptions are made about the size, shape, and orientation of the particles, this method is unbiased. Obviously, the section thickness must be measured very precisely. In an extensive study (De Groot DMG, submitted), it was found that the results obtained with different techniques—i.e., the Small fold method, an interference microscope, and electron microscopic comparison of exposure densities —did not differ significantly.

The method described in this chapter is no longer more laborious than the conventional random section methods for calculating N_V, especially when it is considered that, in the conventional methods, the mean particle size must often be derived from the deconvoluted profile size distribution.[1] Apart from being rather laborious, the different deconvolution methods do not lead to identical estimates. In general, conventional methods may be applied only when absolute certainty exists that the real shape of the structure resembles an idealized shape. In all other cases the method of Cruz-Orive (Ref. 2 and Appendix) is to be preferred. It is always advisable to check first whether large variations in size and shape exist among the particles to be studied. Subsequently, one has to decide which method will be used for deriving the numerical density of the particles of interest. In the case of quantification of synapses, no randomly selected E-PTA- and OsO_4-treated sections should be used when perforated and nonperforated synapses are simultaneously present in the tissue. Random semithin (500-nm) E-PTA sections are very useful for a quick examination of variations in size and shape, but they should not be used for quantification of synapses of different sizes and shapes.

APPENDIX: ESTIMATION OF N_v FROM SERIAL SLABS, by Luis M. Cruz-Orive

Introduction

In Cruz-Orive[2] two methods were developed to estimate the numerical density $N_V \equiv N/V$ of a collection of N disjoint particles of arbitrary size, shape, and orientation contained in a solid Ω of volume V. The methods are based on measurements made on m uniform random blocks of material from Ω. From each block, a number of parallel sections a known distance apart are needed. The following assumptions are made: (1) The sections have zero thickness and (2) there are no truncation effects (that is, every particle fragment in a section is observable).

In this Appendix we readapt the most efficient method, namely method I of Cruz-Orive,[2] with assumptions (1) and (2) relaxed. It turns out that the estimate of N_V remains unaffected by truncation effects.

It must be pointed out, however, that an unbiased estimate of particle "caliper length" or linear projection H perpendicular to the sections is not available in general in the presence of truncation without additional assumptions. What is estimable is just $\tilde{H}$, the "effective" linear projection, which can be interpreted as H minus the linear projection of the truncated part [for an example pertaining to spheres see Cruz-Orive, Ref. 30, Eq. (8.3)]. Fortunately, it is $\tilde{H}$, and not H, that is needed to estimate N_V, as shown below.

The results are given first. Proofs are outlined at the end.

Estimation of N_V

Each of m uniform random blocks from the reference solid Ω is cut into serial sections of a known thickness. A given section toward the middle of the series is taken as the *reference section*. An approximately unbiased estimate of N_V is computed as follows:

$$\text{est}(N_V) = \frac{\sum_{i=1}^{m} N_i n_i^{-1} T_i^{-1} \sum_{j=1}^{n_i} m_{ij}^{-1}}{\sum_{i=1}^{m} A_i} \tag{A.1}$$

where $\text{est}(\theta)$ denotes the estimate of a fixed quantity θ and

- N_V = number of particles per unit volume of reference solid
- m = number of uniform random blocks cut into series
- A_i = reference area, namely the sampled area of the reference section from the ith block ($i = 1, 2, \ldots, m$)
- N_i = total number of particle sections (not just particle "fragments") counted in A_i, according to an unbiased counting rule (see Ref. 2, Fig. 1; and Ref. 14)
- n_i = number of particles in A_i tracked through the series
- T_i = thickness of sections of the ith series
- m_{ij} = number of serial sections from the ith block in which the jth tracked particle appears. Note that j runs from 1 to n_i

Under certain assumptions[5] a good estimate of the standard error of $\text{est}(N_V)$ is

$$\text{SE}\{\text{est}(N_V)\} = \left\{ (m - 1)^{-1} \left[\frac{\sum_{i=1}^{m} y_i^2 x_i^{-1}}{\sum_{i=1}^{m} x_i} - \text{est}(N_V)^2 \right] \right\}^{1/2} \tag{A.2}$$

where we have put $y_i = N_i n_i^{-1} T_i^{-1} \sum_{j=1}^{n_i} m_{ij}^{-1}$, $x_i = A_i$. The coefficient of error in percent is

$$\text{CE}\{\text{est}(N_V)\} = 100 \times \frac{\text{SE}\{\text{est}(N_V)\}}{\text{est}(N_V)} \tag{A.3}$$

Notes:

N1. The orientation of each series may be chosen at will.

N2. The distance between the first and the last sections of a series must be at least twice the longest projection of a particle in a direction perpendicular to the sections.

N3. Instead of using all sections in a series, one out of every two, three, etc. may be used. In this case, T_i must be replaced respectively with $2T_i$, $3T_i$, etc. in Eqs. (A.1) and (A.2).

Outline of a Proof of Eq. (A.1)

A proof of Eq. (A.1) consists of three steps. First, for a uniform random reference section of area A, thickness T, and fixed orientation, hitting Ω, it can be shown that

$$N_V = N_A \, (E\tilde{H} + T)^{-1} \tag{A.4}$$

where N_A = (number of particle sections captured in A)/A and $E(\cdot)$ denotes the true mean value for all particles in Ω. Second, an unbiased estimate of $(E\tilde{H} + T)^{-1}$ is

$$\mathrm{est}(E\tilde{H} + T)^{-1} = n^{-1} \sum_{j=1}^{n} \{(\tilde{H}_j \mid \uparrow) + T\}^{-1} \tag{A.5}$$

where n is analogous to n_i (see above) and $\tilde{H}_j \mid \uparrow$ denotes the effective linear projection of a tracked particle, that is, of a particle actually hit by the reference section (hence the symbol " $\uparrow$ "). Equation (A.5) holds because the probability of hitting and observing a particle is proportional to $\tilde{H} + T$. Third, it can be shown that

$$\mathrm{est}\{(\tilde{H}_j \mid \uparrow) + T)\} = m_j T \tag{A.6}$$

is also unbiased, where m_j is analogous to m_{ij}. Note, however, that $(m_j T)^{-1}$ is only "approximately unbiased" for $\{(\tilde{H}_j \mid \uparrow) + T\}^{-1}$. Hence Eq. (A.1) is approximately unbiased too. Combining Eqs. (A.4)–(A.6) it follows that an approximately unbiased estimate of N_V is

$$\mathrm{est}(N_V) = N_A \cdot (nT)^{-1} \sum_{j=1}^{n} m_j^{-1} \tag{A.7}$$

If m independent blocks with reference areas $A_1, A_2, \ldots, A_m$ are available, the ith N_V estimate should be weighted by $A_i/(A_1 + \cdots + A_m)$. These weights are preferable to the ones proportional to $n_i A_i$ proposed in Cruz-Orive [Ref. 1, Eq. (28)]. Thus, we obtain Eq. (A.1).

From Eqs. (A.5) and (A.6) we see that we can estimate $E\tilde{H}$ but not EH, as pointed out in the Introduction to this Appendix. Fortunately, it is $E\tilde{H}$ that is needed to estimate N_V, not EH [see Eq. (A.4)].

Note Added in Proof

After this chapter was written, H. J. G. Gundersen (J Microsc 143:3–45, 1986, Section 3.3 and Fig. 3.3, left) made a subtle but important remark concerning the unbiasedness of the estimation of particle height via Eq. (A.6), and hence of N_V using Eq. (A.1). The reader is referred to that paper for details. The remark pertains specifically to the situation described in our Note N3, p. 155. If, on the contrary, the height of the sampled particles is estimated by means of consecutive, adjacent sections (as was the case in the present application), then the unbiasedness of Eq. (A.6) remains unaffected.

REFERENCES

1. Weibel ER: Stereological Methods, volume 1, Practical Methods for Biological Morphometry. London: Academic Press, 1979.

2. Cruz-Orive LM: On the estimation of particle number. J Microsc 120:15–27, 1980.
3. Verwer RWH, De Groot DMG: The effect of shape assumptions on the estimation of the numerical density of synapses from thin sections. Prog Brain Res 55:195–203, 1982.
4. De Groot DMG, Bierman EPB: The complex-shaped "perforated" synapse, a problem in quantitative stereology of the brain. J Microsc 131:355–360, 1983.
5. Cruz-Orive LM: Best-linear unbiased estimators for stereology. Biometrics 36:595–605, 1980.
6. Peters A: The fixation of cerebral nervous tissue and the analysis of electron micrographs of the neuropil, with special references to the cerebral cortex. In: Contemporary Research Methods in Neuroanatomy, edited by Nauta WJH, Ebbeson SOE, p. 56. Berlin: Springer, 1970.
7. Small JV: Measurements of section thickness. In: Proceedings 4th European Congress on Electron Microscopy, edited by DS Bocciarelli, volume 1, p. 609. Roma: Tipografia Poliglotta Vaticana, 1968.
8. Smith FH: A photo-electronic phase-measuring microscope. J R Microsc Soc 87:165, 1967.
9. Gillis J, Wibo M: Accurate measurements of the thickness of ultrathin sections by interference microscopy. J Cell Biol 49:947, 1971.
10. Williams MA: Autoradiography and immunocytochemistry. In: Practical Methods in Electron Microscopy, edited by Glauert AM, volume 6. Amsterdam: North-Holland, 1977.
11. Silverman L, Schreiner B, Glick D: Measurement of thickness within sections by quantitative electron microscopy. J Cell Biol 40:768, 1969.
12. Casley-Smith JR, Crocker KWJ: Estimation of section thickness, etc. by quantitative electron microscopy. J Microsc 103:351–368, 1975.
13. Beertsen W, Everts V, Houtkooper JM: Frequency of occurrence and position of cilia in fibroblasts of the periodontal ligament of the mouse incisor. Cell Tissue Res 163:415–431, 1975.
14. Gundersen HJG: Notes on the estimation of the numerical density of arbitrary profiles: the edge effect. J Microsc 111:219–223, 1977.
15. Vrensen G, De Groot D: Phosphotungstic acid staining and the quantitative stereology of synapses. In: Electron Microscopy and Cytochemistry, Proc 2d Int Symp, edited by Wisse E, Daems WT, Molenaar J, van Duyn P, p. 255. Amsterdam: North-Holland, 1974.
16. West MJ, Coleman PD, Wyss UR: A computerized method of determining the number of synaptic contacts in a volume of cerebral cortex. J Microsc 95:277–283, 1972.
17. Greenough WT, West RW, De Voogd TJ: Subsynaptic plate perforations: changes with age and experience in the rat. Science 202:1096–1098, 1978.
18. Vrensen G, Nunes-Cardozo J: Changes in size and shape of synaptic connections after visual training: an ultrastructural approach to synaptic plasticity. Brain Res 218:79–97, 1981.
19. Gray EG: Axosomatic and axodendritic synapses of the cerebral cortex: an electron microscope study. J Anat 93:420–433, 1959.
20. Peters A, Kaiserman-Abramof JR: The small pyramidal neuron of the rat cerebral cortex: the synapses upon the dendritic spines. Z Zellforsch Mikrosk Anat 100:487–505, 1969.
21. Akert K: Dynamic aspects of synaptic ultrastructure. Brain Res 49:511–518, 1973.
22. Cohen RS, Siekevitch P: Form of the postsynaptic density: a serial section study. J Cell Biol 78:36–46, 1978.
23. Dyson SE, Jones DG: Quantitation of terminal parameters and their inter-relationships in maturing central synapses. A perspective for experimental studies. Brain Res 188:43–59, 1980.
24. Vrensen G, Nunes-Cardozo J, Müller L, Van der Want J: The presynaptic grid: a new approach. Brain Res 184:23–40, 1980.
25. Müller L, Pattiselanno A, Vrensen G: The postnatal development of the presynaptic grid in the visual cortex of rabbits and the effect of dark-rearing. Brain Res 205:39–48, 1981.

26. Abercrombie M: Estimation of nuclear population from microtome sections. Anat Rec 94:239–247, 1946.
27. Fullman RL: Measurement of particle sizes in opaque bodies. Trans AIME 197:447–452, 1953.
28. Burry RW, Lasher RS: A quantitative electron microscope study of synapse formation in dispersed cell cultures of rat cerebellum stained either by Os-Ul or by E-PTA. Brain Res 147:1–15, 1978.
29. Mayhew TM: Stereological approach to the study of synapse morphometry with particular regard to estimating number in a volume and on a surface. J Neurocytol 8:121–138, 1979.
30. Cruz-Orive LM: Distribution-free estimation of sphere size distributions from slabs showing overprojection and truncation, with a review of previous methods. J Microsc 131:265–290, 1983.
31. De Groot DMG, Bierman EPB: A critical evaluation of methods for estimating the numerical density of synapses. J Neurosc Meth 18:79–101, 1986.

part 6

outlook and image analysis

This part is meant to serve as a look into the future of the quantification of morphological structures and the objective study of relationships between cells or cellular components. Although morphometric/stereologic studies can be performed virtually without expensive equipment, some type of aid may be advisable if this type of analysis becomes a regular part of the work. This is particularly true for computerization, which is a by-product of all the larger semiautomatic machines, e.g., the computer-assisted tracing system MOP-Videoplan by Kontron, or automatic machines such as the TAS and IBAS systems of Leitz and Zeiss/Kontron, respectively. Real progress will occur in automatic devices, provided they become both cheaper and more sophisticated. This seems to be happening very quickly, and therefore some knowledge of the potential of these systems and techniques seems necessary for morphometrists and stereologists.

Rigaut's chapter treats the impact of image analysis in electron microscopy by describing the development of the systems and giving examples of applications and a look into the future. It should be borne in mind that these machines analyze only two-dimensionally and that it is up to stereology to reveal the three-dimensional structure.

chapter 13

analyzing electron microscopic images by computer: a guided tour

Jean Paul Rigaut

INTRODUCTION

The best image analysis is still most often accomplished by the human eye and a brain prepared, by long education and varied experiences, for complex tasks, some of which require multiple and unpredictable decisions. This is particularly true for morphometry in biological electron microscopy, where the relevant features usually display rather poor contrast and may be drowned in a complicated background. In addition, the specimen may have been altered by drastic preparation and the electron beam itself may have destroyed some information. The difficulties encountered in automated computer analysis of electron images can be insurmountable. Our eyes, however, are sometimes surpassed—for instance, in some studies of molecules, where preliminary image enhancements are often required before seeing anything at all. In such cases, complex image restorations and three-dimensional (3-D) reconstructions are necessary anyway. In other situations, our psychological resistance is surpassed; no one appreciates the tedious measuring of thousands of particles. This does not mean, however, that the computer is always able to perform the required task. In biology and pathology, when results are asked for, manual evaluations using morphometric point and line grids will very often still be preferable. Only some very fortunate cases, in which the features' contrast is excellent or can be sufficiently enhanced, will allow straightforward use of an automated image analyzer. This, however, should not slow down the enthusiasm of the pioneers who devise more and more elaborate algorithms to cope with other situations. The computer will win— is winning—but in some cases it will take many years before the biologist's or the pathologist's call for reliable and fast results in a fully automated way is satisfied. Semiautomated image analyzers allow manual modifications and contourings of features on a monitor screen and still offer the advantage of computer filing of the data, but, as we will see, they are usually not more efficient than simple evaluaton by the eye based on grid countings on micrographs.

The term "image analysis" has many different meanings, especially in electron microscopy: evaluation of imaging conditions, image enhancement, segmentation, 3-D reconstruction from images obtained at different tilts or from serial sections, pattern recognition, and acquisition of numerical data that will often become meaningful only after suitable mathematical treatments (stereological estimations for instance). Many, often highly specialized, techniques use digital spatial frequency filterings (Fourier transforms), mainly in the field of molecular reconstruction, to obtain information on the periodicity and arrangement of structures. Image enhancement methods will be discussed, including spatial frequency filterings, but also "image classification" processes, which tend to reduce the information contained in an image to a few parametric data about sizes, shapes, and compositions of features, through geometric and densitometric operations in the computer. The latter will then usually be of a specialized type: the "image analyzer," as opposed to general-purpose computers, which are more often used for complex molecular studies. The analyzer, used in an automated or semiautomated (manually interactive) way, will digitize an image coming from a TV camera [transmission electron microscopy (TEM), usually with micrographs], or by direct electronic signals [scanning transmission electron microscopy (STEM) or scanning electron microscopy (SEM)].

INSTRUMENTS

History

The first breakthrough was the invention of the Quantimet in 1963.[1] Its basic principle is still one of the essential bases for all image analyzers: scanning of the image in parallel lines by a TV camera, the produced signal being processed by a detector unit. The first-generation commercial image analyzers were of the binary gray-level type, keeping in memory only the features that are revealed by a chosen threshold and width of gray-level. Many additional features were progressively added, such as interactive systems (digitizing table and/or light pen), autofocus and scanning stage for the optical microscope, direct signal input for SEM and STEM, and interfacing to other computers. Then in recent years came the real revolution, which at last offered more possibilities for automated image analysis in electron microscopy: the invention of analyzers that store—and work on—the image with all its gray values, the binarization becoming a late step, preceded by unlimited possibilities in enhancements and filterings. Recent instruments have very few limitations, but they require more computing skill if one wants to use them at an optimal level; special software packages include highly sophisticated interactive image and data treatment "menus." An important achievement was the coupling of an image analyzer to an SEM by direct interfacing.[2]

The story of image analysis in electron microscopy is a short one. The first TEM work seems to have been done by Strang,[3] on steel alloys with the Quantimet. Since then, many significant EM metallographic and molecular studies have been achieved, but there have been very few biological ones compared to the number in optical microscopy.

Image Digitizing

Different paths can lead from the electron beam to the computer, through direct use of the electronic signal (SEM, STEM), or a TV camera, viewing an electron micrograph (TEM, SEM) or a transmission phosphor situated at the base of the column (TEM). In the latter case a lens or a fiber-optic system is used,[4] but the image produced on the phosphor must be sufficiently bright or an image intensifier device must be used (such interfaces have recently been commercialized).

A micrograph can be viewed either as a print or as a negative on an epidiascope attachment. Several special photographic methods may be used to facilitate detection by a TV camera, such as a very high contrast print with a lith-type developer and a high-contrast process film ("posterization") or prints on isodensity films[5]; it is also possible to delineate some objects with a black felt-tip pen or to use correction fluids to suppress unwanted details. Instead of viewing the micrograph with a TV camera, it is possible to use a microdensitometer to produce a discrete digitized version of the photograph.

With SEM instruments, the specimen is scanned by an electron beam and many types of images are produced simultaneously (secondary electron, backscattered electron, absorbed electron, transmission, X-ray, cathodoluminescence, magnetic contrast, selected-area and large-area channeling pattern images) as electric signals which may be processed directly (i.e., without an optical TV camera interface). The computer must be synchronized with the SEM by digital scan generators.[4]

Image Analyzers

Semiautomated instruments, always less expensive, help in many morphometric studies, but potential users should first of all wonder whether their problem would not require simple point or linegrid methods; according to two studies,[6,7] those classical techniques are more efficient when aggregate values are needed. In our view, image analysis should be used in situations where large numbers of individual values are necessary. Semiautomated instruments can also be very helpful when exact object coordinates or contours are required, and one of their pleasant features is the computer data filing. All fully automated analyzers also offer digitizing tablet facilities, and this allows one to study micrographs whose poor contrast would forbid any automated task. Fully automated image analyzers are very expensive and should certainly not be acquired from the start just because one wishes to join the ranks of those who describe biological objects with numbers. However, in our opinion, they may be used in biological electron microscopy in two very different cases: when it can be ascertained beforehand that some practical results will be obtainable without years of effort or when a laboratory wishes to specialize in pioneering new image analysis algorithms for the future (then, time does not count). The fields of application are broader in materials science and in molecular studies.

All automated image analyzers use mainly the optical density ("gray level") of the different features in the image as the usual basis for their recognition, the general process being named detection. The first-generation image analyzers kept in the image memory a specific gray-level range, binarized at entry. The more recent instruments start with complete memorization of the image, often in 256 gray levels (8 bits), and this is highly recommendable in electron microscopy, as pratically all situations will necessitate image enhancements of all possible sorts before any binarization.

SPECIMEN PREPARATION

State-of-the-art histological techniques for EM may be found in numerous books. It must, however, be emphasized that (as in optical microscopy) we should endeavor to develop new methods more adapted to image analysis. The ideal goal is perfect contrast of a specially stained feature, but this is very seldom feasible! It might be useful here to cite some special staining methods that help greatly in some TEM cases. The Feulgen nuclear reaction can be adapted to EM[8] and DNA can be stained by uranyl acetate after hydrolysis.[9] Ruthenium red allows one to visualize cell surface glycosaminoglycans, and this has allowed an image analysis study of the conjunctiva.[10] The strong complexing of osmium to dimethylaminobenzidine (DMAB) in the peroxidase staining technique has been used for image analysis with anti-serum protein antibodies to study cerebral extravasation[11] or with monoclonal anti-T6 antibodies in skin histiocytosis X.[12] Peroxidase has also been used to quantitate monocytic lysosomes.[13]

A specific stain for degenerating nerve terminals and lysosomes[14] has been used for image analysis.[15] In all those cases, better discrimination can be obtained by omitting the traditional uranyl/lead staining steps. EM staining problems in image analysis are reviewed elsewhere.[16]

For SEM staining, some useful reviews can be consulted.[17] Carbon coating must replace gold coating when using backscattered electrons (BSE);

in this case, a new silver technique allows quantitative studies of the inside of whole cells through their surfaces,[18] and nucleolar organizer regions (NORs), nucleoli, and nuclei can be vizualized.

We wish to emphasize the importance of quantitative variations due to histological preparations. For TEM, considerable variations can be recorded, depending mainly on the type and duration of fixation but also on the buffer molarity.

IMAGE PROCESSING METHODS

The numerous and highly sophisticated algorithms, which have been extensively described in the specialized literature, cannot be explained in detail here. Those that apply mainly to molecular reconstruction schemes will be very briefly discussed. The final aim of image computer processing can be either the acquisition of new qualitative information (e.g., molecular structure reconstruction) or the extraction of quantitative data. Image enhancement is an important first step in both cases; in the former, "showing the invisible" is often the main concern, and in the latter, no feature can be evaluated if its contrast does not allow its recognition by the computer. In molecular studies, 3-D reconstruction will be an essential step; in quantitative studies, pattern recognition will be predominant, as no automated measurement is possible without the desired object being isolated beforehand from other features and the background. Of course, semiautomated techniques will always be feasible when full automation fails. The problem of isolating interesting features from complex images sometimes has no solution at present by automated methods. This is often the case in EM, where contrasts are usually poor.

Digital Image Enhancement Methods

The range of possibilities is considerable. The three-dimensional reconstruction of the purple membrane protein molecules, virtually invisible on the original micrographs, is the notable achievement.[19]

Diagnosis of Imaging Conditions

Two different mechanisms are involved in contrast formation[20]: amplitude contrast and phase contrast. The former arises from electrons deflected enough (by very coarse features and/or heavy-metal stain) to be stopped by the objective aperture. Finer structures will still deflect the electrons, but not enough for them to be stopped by the objective. Specimens that produce very small electron deflections and absorb very little energy are known as weak-phase, weak-amplitude objects. For the amplitude effect, the image intensity is related linearly, by a convolution formulation, to the specimen phase distribution. For a weak-phase object, the density at each point of the Fourier transform of the image is equal to the intensity at the corresponding point in the Fourier transform of the phase distribution multiplied by an operator, which is called the transfer function and is related to the microscope parameters. For a weak-phase, weak-amplitude object, the Fourier transform of the amplitude distribution, also multiplied by a transfer function, must be added. Electron microscope transfer functions depend on the wavelength, the defocus, and the coefficient of spherical aberration. In biological practice, very few objects are of the weakly scattering type. In other cases, the image intensity depends on an additional mathematical term and no direct linearity exists between its spatial frequency spectrum and the distributions of the phase and amplitude of the specimen. Computer processing of several micrographs taken at different defocus values can be used in such cases, in a Fourier space context, to estimate the phase and amplitude components.

Radiation damage is the "ultimate hurdle for high resolution biological structure research with the electron microscope".[21] A sequence of images taken with increasing exposures may be used to assess the extent of radiation damage.[20] A large number of images can be obtained from a single minimum-exposure picture and combined for a study of fragile structures.[19]

Aberrations Due to the Image Analysis System

Great care must be taken to choose a suitable resolution.[22] The major criterion, in terms of a number of pixels per unit length, is the distance between the TV points constituting the memorized picture. The spatial resolution of a TV-scanning system can be evaluated by measuring a sharp edge, differentiating the digitized response, and calculating the modulation transfer function (MTF) using the Fourier transformation. The information content of a digitized image depends on the MTF of the whole measurement system.[23]

When studying a micrograph with a TV camera, care must be taken to produce homogeneous illumination and no reflections. A "shading" problem is nevertheless often present; it can still be corrected after storing a correcting matrix (an image of the nonuniform background is subtracted).

The halo error, where some thin perimeters around white or black phases take a grayish value even if the edge contrast of the object is remarkable, is explained by finite resolution effects.

An unacceptably high degree of noise (random voltage variations) will produce a "snow" of fine bright spots. Smaller amounts of noise explain the flickering of small objects on the monitor as they are alternately detected or not by successive scans. This is not to be confounded with the "background noise" in an EM image due to the object's nature itself or to the scattering of deviated electrons.

Geometric distortions of images can arise from TV camera lenses and SEM scanners.

Measurement errors and screen edge problems will be examined later (Data Extraction section).

Image Restoration

This term is used to denote methods for retrieving the object function from an image distorted by instrument aberrations.[21] Restoration is easiest in specimens that display a periodicity, by the use of optical or digital Fourier transforms. It may be known, for instance, that the relevant information, in the coarse features of the image, is obscured by a noise of fine details; most of the unwanted background can be eliminated by taking the Fourier transform of the image and removing its outer part (high spatial frequencies). The inverse Fourier transform reconstitutes the (corrected) image.

Restoration of nonperiodic objects is more problematic.[21] Instrument-related aberrations can be eliminated only if the cause of the defect is well understood and if a given type of defect is always due to the same cause.[20] The latter condition is known as the uniqueness clause. For weakly scattering objects, the simplest restoration method again uses the Fourier transform. Another method uses the information from several images obtained at different defocus values.[24]

Deblurring of an image involves several corrections.[25] Blurring can be understood as a convolution process, and Fourier transforms can again be used to obtain deblurring. The methods are rather complex,[26] although simple weighted subtraction from the image of its Laplace transform (see below) can sometimes be quite efficient,[27] and removal of blur caused by uniform linear motion by Fourier domain methods can be relatively easy.

Noise filterings have been proposed, again in a Fourier context.[26] The optimal linear filter may be the matched filter[29] when the noise is assumed to be of the white Gaussian type. When the Fourier transform of the random noise image can be expressed as a rational function, the optimal filter corresponds to a weighted mixture of the signal and its Laplace transform.[26,27] High- and low-pass filters can be used, even for nonperiodic objects,[30] as well as correlation functions (see below).[31]

Averaging and Alignment Methods

These methods represent a logical way to distinguish the incidental from the reproducible[21] and are used in molecular or crystal reconstruction studies. Local Fourier patterns can help in aligning serial sections.[32] The method of spatial averaging is based on the fact that the Fourier transform of a periodic object is concentrated in narrow zones.[21] The cross-correlation function can be used for the alignment of several images of the same object.[30] Various correlation functions can be applied to answer the question of orientations and regularities in image structures[31] and to reconstruct nonperiodic objects[32]; they are also useful in noise reduction.[31]

Contrast Enhancement Methods

The first type is again based on Fourier transforms and convolutions (frequency domain filterings). The principle is simple: if, for instance, the relevant information is contained in the fine detail of the image and is hidden by coarse features, a high-pass convolution filter is used; the Fourier transform of the original is multiplied by the filter. Then the filtered image is obtained by computing the inverse Fourier transform. Edge accentuation can be achieved in this way. The problem of selecting the filter is addressed by Hawkins.[26]

Spatial domain filterings are convolutions based on local gray-level operations: the output density at each image point is computed from the input one and those of a certain number of its neighbors by centering a filter matrix on each successive pixel. Then the new value of the central pixel is the weighted sum of the point values, the weights being the matrix coefficients. Some recent commercially available image analysis software proposes contour enhancement filters of this type.

In spatial domain "point mapping operations," the output density of each point is calculated from the input one according to a function, but independently of neighborhood values. The usual motivation is to use more efficiently the available gray scale of the display device. The density histogram of the image is used to compute the parameters of the mapping function, in such a way that the output picture histogram becomes as close to uniform as possible (histogram linearization). The small density variations occurring at the most frequent gray-level values are spread out, becoming more visible, but this happens at the expense of definition in parts of the image where some values occur less frequently. The normalization of the gray-level histogram is another useful contrast enhancement operation, through which the useful range of gray levels is remapped to the maximum available range.

Image Smoothing Methods

These methods are used mostly to attenuate some undesirable effects due to poor sampling or transmission, but they also allow simplification of some of the information content of the image, allowing more straightforward subsequent segmentation. In the frequency domain, low-pass filtering can achieve blurring by attenuating a specified range of high-frequency components in the Fourier transform of an image. Neighborhood averaging results in low-pass filtering in the spatial domain. In median filtering, the gray level of the pixels included in the position of the gliding matrix are ranked in increasing order. A parameter (rank) then decides which value in the series will replace the original one of the central pixel. If the rank is equal to half of the square of the size plus one, the filter is a "true" median and serves as a low-pass filter for noise reduction. It does not, in contrast to average filters, blur sharp edges and contours. Other ranks can be used for special effects (Fig. 1).

In practice, two situations must be distinguished: those in which image reconstruction is primordial (e.g., molecular studies) and the frequency domain is paramount, and those arising in cell biology, which use spatial do-

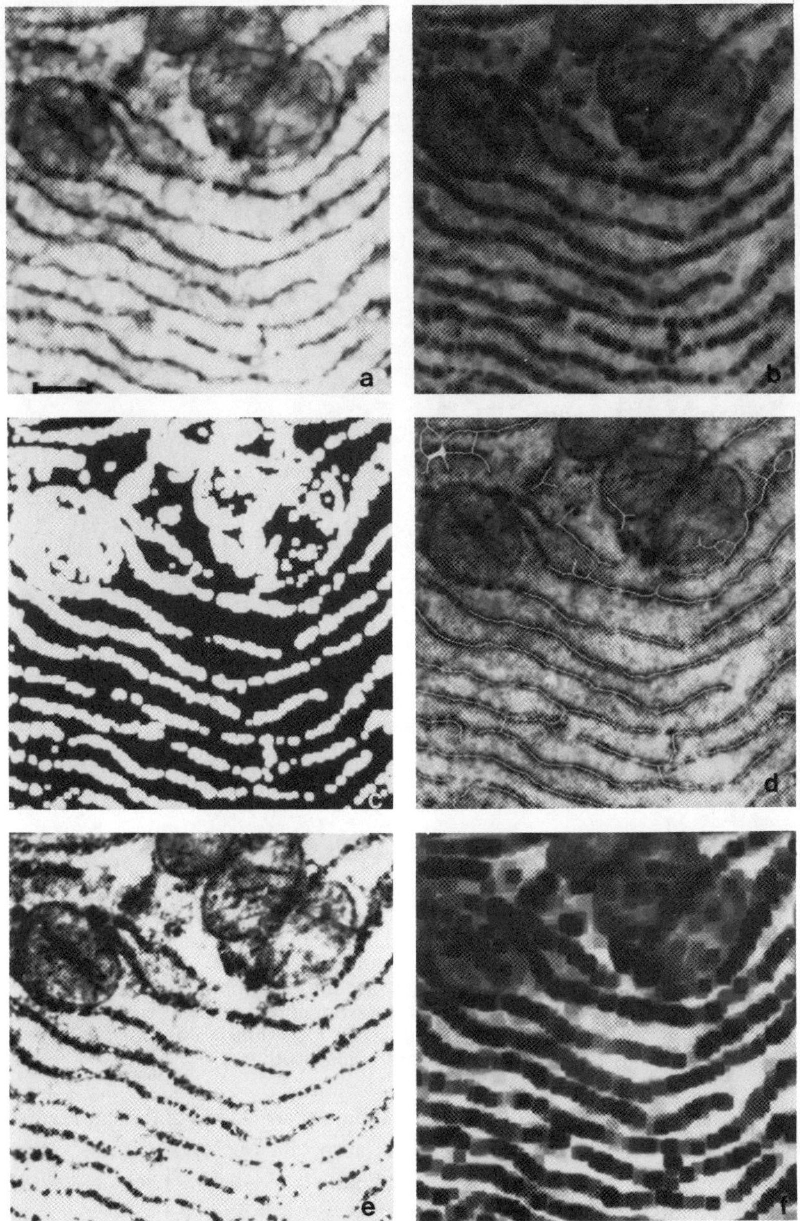

FIGURE 1. Automated image analysis (IBAS, Kontron, FRG) of an EM section of a rat hepatocyte after glucose-6-phosphatase reaction. Micrograph (Dr. A. Reith), ×30,000; final calibration, 5.2 nm/pixel (bar, 0.4 μm). (a) Original TV camera (memorized) image on the monitor. (b) After median filtering (7 × 7 pixel matrix, rank 4); increased detectability of reticulum. (c) Binarized image (gray-level range, 0–112); the interrupted reticulum is then reconnected by a closing operation. Comparisons of images before and after openings allow the elimination of mitochondria. (d) Final image, skeleton of reticulum superimposed on the original image; note some errors (e.g., some skeleton lines in mitochondria). (e) Image obtained from the original one with another median filter (7 × 7, rank 40); (f) Same comment, but (15 × 15, rank 4); with these two filters it would not be possible to obtain a correct final result.

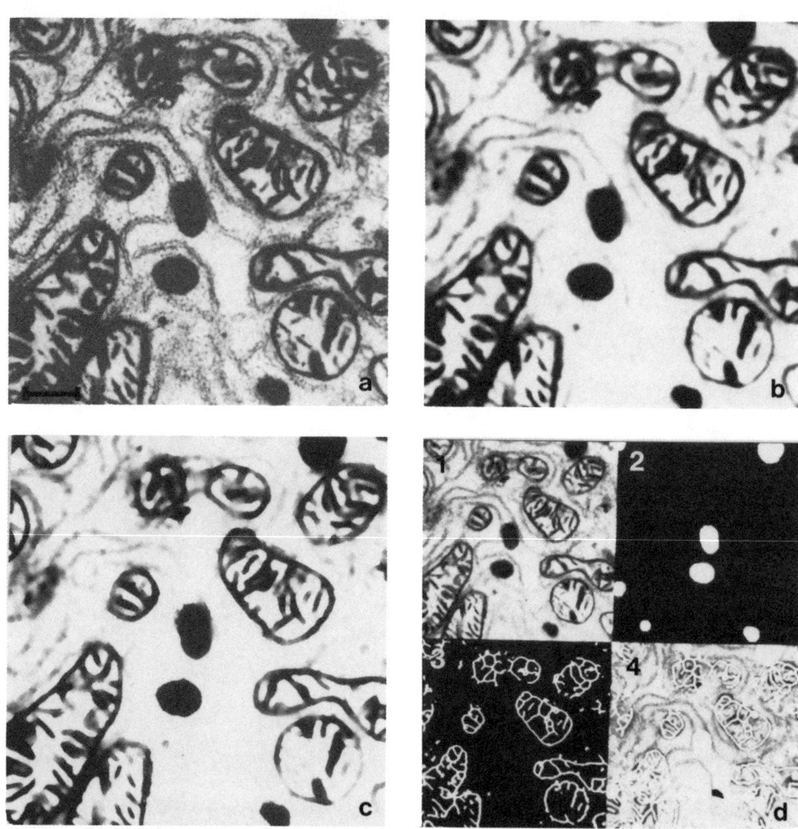

FIGURE 2. Automated image analysis (IBAS) of an EM section of a mouse hepatocyte after peroxidase (catalase) reaction. Micrograph (Dr. A. Reith), ×30,000; final calibration, 5.2 nm/pixel (bar, 0.4 μm). (a) Original TV camera (memorized) image on the monitor. (b) After Gaussian filtering (11 × 11 pixel matrix) followed by high-pass filtering (50 × 50 matrix, unsharp masking method). (c) After edge-sharpening (Kontron "delineation" algorithm; Sobel high-pass filtering followed by halo elimination by masking). (d) Final result: 1, original image; 2, lysosomes; 3, skeletons of mitochondria; 4, 2 and 3 superimposed on 1.

main methods. In the former case, the techniques are varied and complex but well codified; in the latter, each case is a special one and trial and error usually represents the only possibility (several successive filterings are often used). An example of image enhancement is given in Fig. 2.

Digital Image Classification Methods

All the techniques reviewed here consist of reducing the information contained in the image to a few qualitative (e.g., "are such objects present?") and/or quantitative (e.g., "what is the mean area of those objects?") results. The image must often be segmented, and geometric, topological, or textural patterns recognized, before any extraction of quantitative data. Many methods use various operations derived from "mathematical morphology," which is not an image processing technique in itself but offers a sound theoretical basis for a great number of modifications of sets of pixels, in binary or even in multi-gray-level images.

Mathematical Morphology

We cannot do more here than barely evoke such a theory, pioneered by Matheron and Serra.[33] The basic principle is the "hit-or-miss transformation": the center of a structuring element (small geometrically defined object) is set sequentially on each pixel of a binary image. The "erosion of the feature by the structuring element" is effected by removing all the pixels that do not satisfy the condition that all the other parts (points) of the structuring element must fall inside the feature. The opposite (or the same, but of the background) is a dilation of the feature. An opening is an erosion followed by dilation, and a closing is the opposite.

An opening on a whole field produces an image "cleaned" of its smaller features, as they do not reappear after having disappeared during the erosion. A closing produces a smoothing of wavy feature contours and closes small holes in features. Structuring elements can be chosen to be as close as possible to a circle (hexagon for the texture analysis system [TAS] hexagonal lattice, octagon for a square lattice), or oriented (lines or square) to reveal anisotropic feature tendencies.

Several routinely used algorithms allow complicated image segmentations (e.g., skeletons, gradient extractions, separations). The spatial distribution of features can be studied by the covariance function (the number of particles versus the number of dilations in a binary image). In recent years, openings and closings on multi-gray-level images considered as functions have appeared (e.g., the rolling ball transform[34]).

Automated Image Segmentation

This operation consists of separating different component features in the image, for instance, isolating a given object from all the others before measuring it. Very often, especially in EM, several subsequent operations of quite different types must be used. Detecting the edge of an object is the best way, of course, to detect the object itself.

Edge and contour contrast enhancement by frequency domain filtering. The relationship between Fourier transforms and edge enhancement is that high spatial frequencies are associated with sharp changes in intensity. High-pass filtering can therefore be used to enhance edges.

Edge and contour contrast enhancement by spatial domain filtering. Spatial filterings by sliding-window matrices[35] can be used to accentuate edges and contours. Contour enhancement or gradient filters can be used (the latter in the x or y direction). The Laplace operator is based on partial second derivatives in the x and y directions.

Laplacian filtering enhances edges[36] but also noise. The Sobel[37] edge detector filter should be preceded by a low-pass one. The Roberts[38] filter produces thinner edges than the Sobel but is sensitive to noise. Other algorithms using local operations have been proposed for mitochondria in EM.[39,40] Compass gradient operators obey the same principle, but applied in all eight possible directions. The performances of several filters have been compared.[41]

Edge and contour extraction. Edge contrast enhancement methods are not usually sufficient to obtain edges isolated from any other feature. Edge extraction (detection) algorithms help, of course, in isolating bounded features (particles or their sections, i.e., profiles).

Some methods are based on automated thresholding, i.e., the selection of gray-level ranges from information offered by the gray-level histogram of the whole image. In most models, any transition from a pixel whose gray level is on one side of the threshold, defined as a valley in the histogram, to an adjacent pixel whose gray level is on the other side can define an edge or a contour location. This, of course, is never perfect in biology, where some edges will be interrupted or ill-defined and other features will create many zones of microgradients that will be totally irrelevant to the study. In optical microscopic (OM) cytology, the method has been improved by smoothing the histogram and comparing it to the original, determining differential valleys by a figure-of-merit test.[42] Two histograms of different natures can be used (gray level and gradient magnitude) and the histogram maxima serve as global threshold segmentation values; this has been applied to OM cytology.[43] A third histogram (compactness) can be added.[44] The variation of boundary lengths as a function of the threshold has been used in OM cytology to segment images.[45]

Local thresholding can also be used. Pixel gray-value neighborhoods can be studied[46] or, again as in global thresholding, two-dimensional histograms of gray level versus gradient magnitude.[47]

Locally adaptive thresholds[48] can be obtained by calculating at each point an "edge activity index," defined as the ratio of the maximum gradient magnitude at this point to the average magnitude of gradients in eight directions. The gradient image can also be compared to a blurred version of the original image, obtained by low-pass filtering.[48]

Sliding-window matrices can again be used, with much more sophisticated algorithms and using nonlinear operators. Rosenfeld and Thurston[49] offered two solutions. The first method takes the products of the differences between the average gray levels of pairs of neighborhoods of all sizes. The second method determines a largest size neighborhood (varying from pixel to pixel). Abmayr et al.[50] proposed in OM cytology a nonlinear gradient algorithm followed by maximization of the joint probabilities of pixels considering their spatial relationships and connectivity. Baky et al.[51] used a gradient minimax derivative statistic for locating nuclei in OM cytology. Complicated operators adapted to the detection of mitochondrial membranes in EM have been proposed.[40]

A method derived from mathematical morphology, described by Beucher,[52] also searches for gradients. It uses iterative gray-level windows with dilations of the successive binarized images and intersections and unions between them. The set of pixels whose gradient magnitude is bigger than a given value (g) is the union of all images obtained at successive gray-level windows over the whole available range. The contours obtained with this algorithm are of variable width and are not always closed, even when the objects in the image are visually discrete. Beucher suggests using the entire gradient magnitude function represented by the sequence of previously obtained sets, at decreasing values of g, but noise then arises.

Much more sophisticated edge detection algorithms, too complex to be explained in detail here, have been proposed. Many of them bear some resemblance to pattern recognition schemes (see below), by using, for instance, dynamic programming[53] for multistage maximization of a merit function. Figures of merit have been proposed in several works since the pioneering one of Montanari.[54] The first group of such methods are heuristic, like all those already discussed (i.e., they are not associated with any formal model of an edge). They are usually based on sequential detection. Several algorithms have been proposed.[54,55] Guidance (context) information can be added.[56] A second group of algorithms is nonheuristic (i.e., based on formal models of an edge). Griffith's method[57] is based on determining whether a region does or does not contain an edge or a contour. Hueckel's method[58] searches for the element that best fits the intensities observed in a region.

Interesting surveys of edge (enhancement and/or) detection methods can be found.[48,59,60] Some successes have been obtained in OM cytology. Those methods are still in their infancy in EM, but the pioneering work of Kemmer and Voss[40] on mitochondria in EM clearly points to the future.

Contrast peak extraction. Instead of detecting gradient zones it is possible to detect, in a gray-tone image, "mountains" that display sharp peaks. Another method based on mathematical morphology is the "top-hat transformation"[61]: a high-contrast peak is detected by "cutting" (thresholding) it at two gray levels; the higher level is retained if the lower is not too large. As not all peaks have the same height, the operation is repeated at various gray levels. The union of all the successive images gives a binary mask for all contrast peaks of different "altitudes."

Skeletonization. The idea of skeletons can be understood in terms of the concepts of mathematical morphology.[62] A good intuitive explanation is that of the "grass fire": imagine the object as grass and start a fire on all boundary points simultaneously; if the object is circular, the fire will progress until the grass left represents one ultimate central point; if the object is not circular, at some points the advancing front of fire will intersect on a line the fire front from some other region; such line intersections define the skeleton.

Watershed extraction. This method,[63] which also derives from mathematical morphology, consists of detecting the set of all points on the surface of a gray-tone function (this can be understood as a chain of mountains) that represent minima, resembling the boundaries of geographic water catchment basins.

Separation between stubbornly overlapping features. Very often, the final segmentation obtained through any of the methods already examined yields images in which some theoretically discrete objects touch or, worse, overlap. The simplest method, "multiple gray-level analysis,"[64] uses multiple analyses of each field at increasing gray-level thresholds. The algorithm compares each successive binary picture with the previous one, and any objects that are seen to have merged are identified. The binary images of those objects are recorded as they appeared one step before the merger, obviously slightly smaller than their real size. This algorithm can therefore only be used for counting features. Several methods for tracing separation lines between overlapping objects have been proposed. The first type of algorithm[65] is complex and very accurate; it uses a heuristic search paradigm (a graph traversal procedure that may produce paths minimizing a cost function). The second method[66] is based on mathematical morphology.

Pattern recognition. A good part of the research that has been done on pattern recognition has dealt with writing or speech analysis. At least in the former case, the problem seems simpler than the almost insurmountably complex ones encountered in biology and even more in EM, where the relevant features are often buried in the background. And yet, after more than 30 years of intensive computer research, "reading machines" are still not as good as humans at recognizing freely hand-printed numerals 0 to 9! As pointed out by Ullmann,[67] the human eye uses context information all the time, i.e., it cannot find the edge of an object without first recognizing the nature of the latter. Pattern recognition schemes cannot be explained in detail here. Preceded by image enhancement, they include two stages[68]: property extraction, to obtain a pattern representation of a lower dimensionality than the original image, and recognition. Decision functions are applied to both stages; in the second one, the pattern is assigned to one of several preselected classes. Pattern recognition needs a good deal of mathematical and computer knowledge. Two approaches may be defined[69]: in the statistical decision-theoretic approach, a set of typical features is extracted from the patterns. Each of the latter is represented by a vector. Each pattern is usually recognized by partitioning the feature space, which is often transformed into lower-dimensional spaces. Fourier, Walsh,[70] Hadamard,[71] and Haar[72] transforms have been used for generating pattern features. Karhunen-Loève[73] and Hotelling[28] transforms have been used to reduce the dimensionality of feature space. The most common approach for feature selection is to define an information or distance measure related to the bounds of the misrecognition probability. The failure of linearized statistical decision theory to produce satisfactory classification results has led to the use of nonlinear decision networks: the image is subdivided into n regions and the set of n subimages becomes the data used in designing n classification functions ("features"). Applying these features to the image data produces n-dimensional vectors, which, in turn, serve in the design of a single higher-level (final) decision function.[74]

Self-adaptive systems[75] can themselves modify the rules for associating zones of the decision space to classes.[26] The second approach is the syntactic one. Each pattern is expressed as a composite of its components ("pattern primitives"). Each pattern is recognized by parsing the pattern, using predefined rules concerning its structure. Special "grammars" (such as web, graph, tree, and shape grammars) are generated to describe the patterns. The recognition itself is usually done by matching an input pattern with "sentences" that represent each reference pattern ("template matching"). Probabilistic (stochastic) grammars can be used: a "sentence" can then enter two pattern "grammars" and the ambiguity will have to be resolved.

Many interesting review papers on pattern recognition have appeared.[37,69,72,76-80] It must be noted, however, that no real pattern recognition work has yet been published in biological EM. In our opinion, the future is, however difficult it will be, in artificial intelligence-related schemes for the recognition of EM features.

Textural recognition. As we shall see later ("Data Treatment" section), textural information can be extracted from a multi-gray-level image by several methods. It can also, although this is still difficult in most cases, be used for image segmentation and recognition.[81]

Fidelity and error criteria for image segmentation. The best test is always . . . the human eye! No computer can yet match its image segmentation ability or its knowledge of the nature of image components. It might be useful, however, to use sometimes statistical tests to compare the accuracy of detection and segmentation. Yasnoff et al.[82] proposed a "generalized quantitative error measure," based on the comparison of both pixel class proportions and spatial distributions of test and reference segmentations.

Semiautomated Methods

The possibilities of modifying an image manually are numerous and most of them exist on all commercial instruments. Two devices exist: either a light pen, used directly on the image display monitor screen, or a cursor, used on the digitized tablet; in both cases, a bright spot shows on the image the position of the "target." Most notably, it is possible to modify some gray levels, fill some interrupted contours, erase undesirable features, or contour any feature and erase all surroundings. The gray-level thresholding can be chosen interactively by a coordinate system with the tablet cursor in some instruments. The digitizer tablet also allows one to trace contours or take any coordinates directly from micrographs placed on it.

Most commercial image analyzers offer a "menu field," that is, a set of ready-made image transformations and data extraction routines that can be chosen by simply using the cursor or light pen as a "mouse" (i.e., tapping with it in the desired sector of the menu).

Data Extraction

Once the relevant features have been isolated, the computer must draw out some quantitative data. Object counting is relatively straightforward, but some precautions must be respected to eliminate any bias due to field-edge errors (objects cut by the frame edges). Several schemes have been proposed, the simplest being to use a virtual frame smaller than the image memory's one and count only objects that do not overlap two of the sides of the frame; this method of "forbidden lines" can be extended to any object shape.[83] The problem of frame-edge errors can also be treated by the Miles method,[84] as generalized by Lantuéjoul,[85] or other statistical corrections.[86] Rigaut has proposed a method for correcting erroneous boundary length measurements in image analyzers that include the edge intersection lengths in their measurements.[87] Perimeter measurement systems are different from one analyzer to another and their accuracies may vary.[88] Other measurement conditions— and possible errors due to the instrument—are discussed by Cole.[22]

Most recent image analyzers offer the software corresponding to a wealth of different measurements. Among those, one that can be particularly useful is the extraction (by using inertia moments) of the positions and lengths of major and minor axes of elliptical objects (as in the IBAS); this eliminates the problem of having to calculate those axes by second-degree equations from area and perimeter[89] with an imprecision due to the resolution-sensitivity of the latter and the approximation of the (integral) formulation of an ellipse's perimeter. Other useful parameters are Feret projections on any direction and coordinates of centroids.

Scanning Electron Microscope Images

As already seen, with SEM the specimen is scanned by an electron beam and electronic signals of many types can be analyzed simultaneously. Secondary electron images are, of course, most commonly used. Backscattered electron images have started to be used.[90] They show heavy-metal inclusions in a striking way. Our own work in cooperation with a team in Oslo[91] has shown the feasibility of analyzing the interior of intact carbon-coated cells; nucleus, nucleoli, and NORs can be made visible by a special en bloc silver staining procedure,[18] as shown in Fig. 3. The glass cathodoluminescence mode[92] is a kind of STEM with no energy filtering, which shows variations in dry mass distribution. Simultaneous image analysis and ''microchemical'' analysis by the X-ray mode (with energy-dispersive spectroscopy, EDS) is possible.[93]

The computer can be used for several purposes in SEM: automated optimization of instrument setting parameters, image enhancement and restoration,[94] and, of course, although this is not the easiest part, data extraction.

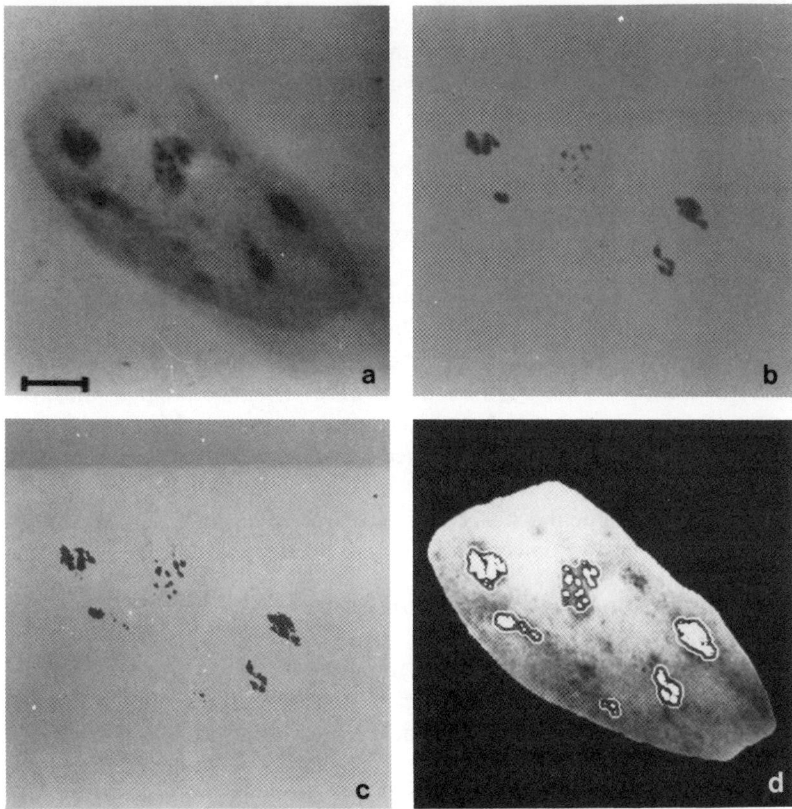

FIGURE 3. Automated image analysis (IBAS) of an SEM micrograph, using the backscattered electron (BSE) mode, on a C3H10T½ fibroblast transformed by benzo[a]pyrene in culture. The cell was whole, coated with carbon, and our modified ''AgNOR'' staining (see text) was used. Micrograph (Dr. A. Reith), ×6660; final calibration, 33.3 nm/pixel (bar, 2.5 μm). (a) Original TV camera (memorized) image, on the monitor, showing the nucleus and stained nucleolar organizer regions (NORs). (b) After high-pass filtering (see Fig. 3). (c) After contour enhancement (7 × 7 pixel matrix). (d) Final result; NORs and nucleoli (contoured automatically) in isolated nucleus.

Some results have been obtained, but for the usually well-contrasted features offered by materials science. We believe nevertheless that an important future exists for SEM computer studies in biology. It might be useful to consult some recent review papers.[95,96]

To Learn More

The interested reader can easily be overwhelmed by the wealth of the literature. For image processing methods, the reader is referred to some major books.[28,33,69,70,76,80,97,98] The International Society for Stereology regularly offers lists of references in its newsletters.

DATA TREATMENT

The reader should be offered a list of the main data treatment methods, with some references. Stereological models and sampling methods will not be discussed here.

Three-Dimensional Reconstruction from Serial Slides

The principle is simple: the contour traces of the sectioned object(s) are memorized for each slide; then a computer program "reconstructs" the object by joining together homologous points chosen on each section trace. In fact, the algorithms are often quite complex. On each outline, the program must decide what will be the minimum number of points to memorize and this is usually dependent on the waviness of the trace. Then the program must define the tiling system. Optimal methods have been proposed.[99] Some superb graphic representations of reconstructed synapses, complete with shading, have been presented.[100]

Mitochondrial structure has been reconstructed in three dimensions.[101] Some other applications in cell biology have been presented.[102,103]

Three-Dimensional Reconstruction from Multitilt Images in Transmission Electron Microscopy

The algorithms proposed to reconstruct an object from several of its projections are extremely complex and this field represents a specialty in itself. An interesting review paper has appeared.[104] The most successful attempts have been made on particular objects having relatively high symmetries. The reconstructed objects were complex protein molecules in practically all cases. For objects of no known symmetry, even more refined methods are necessary. Much attention has been focused on mathematical models, based generally on Fourier transforms.[104] The principle of Fourier 3-D reconstruction is simple (projection theorem): the Fourier transform of a 2-D projection of an object is a "central section" or plane of the 3-D Fourier transform of the object. The origin of each central section corresponds to the origin (0, 0, 0) of the 3-D transform if the projections have a common origin. A certain number of projections are required to obtain the 3-D transform of the object; the corresponding central sections are computed. An attempt is then made to interpolate the unknown values of the full Fourier transform from those on the central sections. A reverse Fourier transform then provides an estimate of the object's 3-D structure. The phase origin of each central section must be determined before any inverse transform can be computed—and this is very difficult for objects that have no symmetry.

Three-Dimensional Reconstruction from Multitilt Images in Scanning Electron Microscopy

There have not yet been any real total 3-D reconstructions made from objects observed in SEM, although the evaluation of the surface of gallbladders[105] is not very far from that final goal.

Many problems will have to be solved before images at different tilts can be used for a full object reconstruction. Using secondary electrons, several types of methods can be used to extract 3-D parameters. The oldest[106] involves shadowing the specimen with metal vapor sprayed from a known angle of incidence and then measuring the length of the shadows cast by the structures. Horn[107] has proposed a sophisticated method using a measure of the intensity of the reflected light. The other methods make use of ''stereo pairs,'' that is, of two images taken at different tilts. In stereophotogrammetry, a special instrument (a mirror stereometer) serves to view the stereo pair and the operator can determine the relative lateral displacement (parallax) of pairs of points from one photograph to the other[108,109]; the mensuration can also be carried out by applying grids on the photographs,[110] and this method can be improved by using a digitizer tablet[111] and a microcomputer.[105] The accuracy of the z coordinates is better than 5% of the maximum height difference of a profile.[105] Mathematical formulations for the extrapolation to three dimensions have been proposed by Lane[112] and Hilliard.[113]

Shape and Pattern Parameters

Except in mathematically definable cases, there is at present no way to define shape unambiguously by any parameter(s). A few ''shape factors'' have a stereological basis,[114] but it is difficult to predict their variations according to changes in (nonmathematically definable) shapes. Most others have a purely 2-D basis and are even quite sensitive to resolution when the perimeter is used.[115] The elliptical eccentricity, or the regularity index (defined, for instance, as the coefficient of variation of the distance between any point on the boundary and the gravity center), and the mean bending energy[116] can be used. Many methods have been proposed for shape classification[117] but none is quite satisfactory.

Mathematical morphology can also serve as a basis in some shape comparison schemes.[33,118,119] Anisotropy can be assessed.[120] Finally, it must be emphasized that boundary lengths are very sensitive to resolution; using fractal theory,[121,122] it is possible to study an object boundary's roughness.[123,124]

Textural Analysis

It is possible to "characterize" the textural composition of an image by several methods. The data here are the gray values and coordinates of the pixels.

The most modern method, based on an early paper by Julesz,[125] allows the use of second-order statistics, with stochastic transition probabilities: for any two gray-levels i and j, the transition probability $p(i, j)$ tells how often i and j occur in adjacent (first neighbor) positions on horizontal scan lines. The gray-level co-occurrence matrix lends itself to numerous image classification algorithms[126] and Markov texture parameters.[127] This method has been used in pioneer works for the diagnosis by OM image analysis of cervical cancer[127] and has taken its place in many image analysis schemes since. It does not seem to have been used yet in EM cell biology.

PRACTICAL ACHIEVEMENTS

It must be emphasized that practical results in EM-based automated image analysis (AIA) are scarce (see an earlier review by Rigaut[128]) in biology and pathology, if one excepts molecular reconstructions, which will not be treated here.

Transmission Electron Microscopy

Cell Nuclei

A simple form factor using the nuclear area and perimeter (measured after manual digitizer-tablet contour tracing) can help in differentiating mycosis fungoides from benign infiltrates on skin biopsies,[129] but with 3% false positive and 50% false negative rates (in early cases). Simple scoring of the number of sharply angled nuclear invaginations confirmed the diagnosis of mycosis fungoides in 100% of cases in another study.[130]

Some remarkable results were obtained[131] on unstained metaphase chromosomes after contrast enhancement; structural components in a small size range (25–56 nm) became visible.

Compared to the numerous studies that have used chromatin texture parameters in OM, those undertaken in EM are scarce. However, they have offered some new information on higher-order conformations of chromatin in *Xenopus*[132] and on the 3-D structure of rat DNA.[133] Chromatin morphometric and/or textural parameters are useful in plant taxonomic studies[134] and in human pathology.[135]

The composition of nucleoli from jejunal columnar cells has been studied in EM.[136]

Finally, nuclear criteria, when associated with other parameters (e.g., mitochondria, lysosomes), allow one to distinguish T from non-T lymphocytes in EM.[137]

Various Cell Organelles

Two teams have proposed algorithms for the AIA of mitochondria[39,40] and changes in internal structure have been monitored in this way in pathology.[138] These algorithms are far from perfect, however, and much progress will be required before computers can really replace morphometric grids for stereological evaluations on mitochondria.

Lysosomes can be studied by AIA[13] after cytochemical staining of their peroxidase content; it was found that monocyte-derived macrophages have a smaller lysosome population than blood monocytes. Peroxisomes from rat and mouse liver and kidney have been studied by AIA and their size distribution estimated by a stereological sphere unfolding algorithm.[139] Cytoplasmic granules are sometimes dense enough to allow AIA measurements. Secretion granules of the anterior pituitary[140] and leukocyte granules[23] have been studied. The latter could be useful in the diagnosis of some forms of leukemia. Human placental coated vesicles have been studied.[141]

Miscellaneous

In ophthalmology, ruthenium red staining allows the quantitation of surface mucins in the conjunctiva.[10] In sclerocornea, the stromal collagenous fibril's diameters have been measured.[142] In the nervous system, a specific stain allows the study of degenerating terminals in brain.[15] The innervation of the cat's canine tooth has been described.[143] The brain capillary network can be mapped.[144] Changes, due to preparative procedures, in brain volume[144,145] and nerve fiber diameters[146] have been reported. EM studies of muscle structure have been made possible by AIA, and this had made it possible to establish the number of strands of thick filaments[147] and to start a 3-D reconstruction of the molecular architecture of the A band.[148] The membranelike structure of the ceramide trihexoside inclusions in the synovial tissue observed in Fabry's disease has been demonstrated by image processing.[149] The diameters of collagen fibers in skin can be measured.[150] Finally, the cytoplasmic density of the so-called light cells of the parathyroid has been evaluated[151]; those cells seem to be artifactually produced by fixation procedures.

Immunocytochemistry

Great hopes are allowed here, as some labelings are dense and contrasted enough to allow AIA quantitations. Some difficulties arise, however, as the real problem is the interpretation in quantitative terms of the immunoenzymatic deposits. The probabilities related to the labelings must be evaluated.[152] The ferritin labeling density for zymogen granules seems to be proportional to the amount of antigens[153]; this "density" should, however, be defined more precisely in most cases, as it covers two distinct notions: the labeling intensity and the labeling area—but then, where is the latter's limit? Well-limited granules[153] are favorable cases. Gold labeling should represent progress,[152] as the gold grains have known dimensions. We have succeeded[128,154] in separating by AIA, using image enhancement and pattern recognition, peroxidase splotches (transferrin) from ferritin spots (albumin) on the same micrograph after double labeling. Peroxidase labeling has also been used to study extravasation of serum proteins in experimental brain tumors[11] and to locate Langerhans cells in cutaneous histiocytosis X, with monoclonal anti-T6 antibodies.[12] Finally, the size distribution of cell surface anionic sites has been studied by cationic ferritin binding.[155]

Three-Dimensional Reconstruction from Serial Sections

Most studies deal with neurons,[156–158] but mitochondria[101] have also been reconstructed.

SEM and STEM

Very few studies have been made by AIA in SEM, at least in biology. In materials science, interesting results have been obtained for many types of particles, inclusions, and fractures of metal alloys.[96]

In biology, stereometry and photogrammetry have been used in only (to our knowledge) two works to obtain quantitative data by AIA: estimation of the height of gallbladder surface wrinkles[105] and of the cell volume and surface area of macrophages.[109]

The rich concretions in the kidney of marine bivalve mollusks have been studied in SEM, using energy-dispersive X-ray analysis (EDXA).[159] The growth of calcium oxalate crystals in kidney stones has been studied.[160] EDXA has also been used to characterize particles extracted from human lungs after occupational exposures.[161]

An early work in materials science emphasized the potential of back-scattered electron images for quantitation in SEM.[90] With the help of a specific silver staining method for NORs,[18] which can be modified to contrast the nucleoli or the whole nucleus, and with whole cells coated with carbon, we have used BSE to penetrate the cells and show those structures (Fig. 3). The combination of secondary electrons and BSE permits study of the surface and the nucleus of the same cell. We have used AIA to show the correlation between plasma membrane and nucleolar parameters when comparing control and carcinogen-transformed cells in culture.[91] A quantitative approach for the estimation of the number of microvilli and evaluation of surface areas on SEM micrographs of nasal epithelia has been proposed.[162]

STEM has been used rather extensively in molecular studies. In cell biology, it has allowed the study of human hepatocellular peroxisomes,[163] including evaluations in pathological cases.

CONCLUSION

The short history of image analysis in biometric EM has not yet led to any truly valuable breakthrough in practical terms, if one excepts biomolecular reconstructions, where some striking results have been obtained. As stated by Weibel[152] for biometry in general, to the question "where are we?" one can answer "not as far as we could be." Considering the availability of ever more sophisticated hardware and software, this must have some explanation. The reasons are not hard to guess: EM images make very difficult the isolation of any discrete object or structure—and most image enhancement and classification algorithms are recent and still at a sort of pioneering stage.

Any potential user of an image analyzer must beware of the fascination exerted by the "toy," which should never mask the "game"[164]—the need for relevant results. We should guard against such fetishism.[165] As has been noted by Weibel,[166] "in the same way as no biochemistry is yet done by the fact of possessing and using an automated spectrophotometer, no stereology is yet done by the fact of possessing and making work an automated image analyser."

The model constitutes the rules of the game. Obtaining data (measurements) without a model through which they acquire their value is akin to playing the game of reality without having the real cards in hand.[167] Images do not represent Nature—they are only partial representations of it—and measurements will be of even more limited value; "there is no image of a triangle which may ever be appropriate to the general concept of a triangle."[168] Image analysis per se does not offer any understanding of biological situations; "stereology is the solid base on which automatic image analysis operates. . . . Stereology defines the measureable parameters and details the required mathematics. Automated image analyzers make statistically significant measurements rapidly."[169]

Once the game has been defined and the interpretative model chosen, and only then, the potential user must wonder how to obtain the relevant measurements in the most accurate and efficient way. Semiautomated methods should be used when the contours themselves are required (before 3-D reconstruction from serial sections, notably) or when exact coordinates of any structure must be known. When measuring numerous discrete objects, a classifying ruler can be used[6] but the user may find less tedious a digitizer tablet, which offers the advantage of filing the data directly into the computer. Automated image analysis in biological EM (excepting molecular studies) must be restricted to those very few cases where the features' contrast either is spontaneously sufficient or can be rendered so by adequate staining methods.

In most situations, we shall have to wait if we want results—and not just emotions—before using automated image analysis: "there is still a long way to go."[166] It is still possible to say that "at the present time . . . it is clear that for many problems the traditional point counting approach is unequalled."[140] The human eye-brain system's supremacy in Gestalt-related activities is yet unchallenged. Of course, pioneers must exist, and in their case the choice of the image analyzer is one of the heart and not of any pragmatism toward immediately obtaining results.

Given the present obstacles that stubbornly resist our drive toward full automation, it is not difficult to guess what the future trends will be in image analysis. Artificial intelligence-derived algorithms will irresistibly, but shyly at first, invade our arsenal of methods: when a feature is difficult to isolate, we must turn to pattern recognition, but it is in such situations that the most stubborn problems are encountered.

The not too distant future is discernible, although one has the awkward feeling that it is always postponed: EM images will travel along elaborate networks from the specimen to various computer memories, through multiple transformations and data extractions, before being stored, after complex compressions, in archiving systems. For the moment, when the computer facilitates in any way our biometric work, it is only through additional difficulties—but those, in turn, push us to refine our understanding of structures, shapes, and patterns.

REFERENCES

1. Fisher C: The Metals Research Image Analysing Computer. Loughborough: Particle Size Analysis Conference, 1966.
2. Braggins DW, Gardner GM, Gibbard DW: The application of image analysis techniques to scanning electron microscopy. SEM 1:393–400, 1971.
3. Strang A: Quantitative electron microscopy. Microscope 16:181–187, 1968.
4. Gibbard DW, Crawley JA, Cowham MJ: The application of image analysis techniques in electron microscopy. Proc 25th Anniv Meet EMAG, pp. 158–161. Inst. Physics, 1971.
5. Mraz P, Kapeller K, Cierny G, Stephanek S: The use of equidensity photographic method in electron microscopy. Anat Anz 142:112–117, 1977.
6. Gundersen HJG, Boysen M, Reith A: Comparison of semiautomatic digitizer-tablet and simple point counting performance in morphometry. Virchows Arch B Cell Pathol 37:317–325, 1981.
7. Mathieu O, Cruz-Orive LM, Hoppeler H, Weibel ER: Measuring error and sampling variation in sterology: comparison of the efficiency of various methods for planar image analysis. Stereology 5, Proc 5th Int Congr Stereol. J Microsc 121:75–88, 1981.
8. Moyne G: Feulgen-derived techniques for electron microscopical cytochemistry of DNA. J Ultrastruct Res 45:102–123, 1973.
9. Erenpreisa J: Staining DNA with uranylacetate in hydrolysed ultrathin sections. Acta Histochem 68:22–26, 1981.
10. Lee WR, Murray SB, Williamson J, McKean DL: Human conjunctival surface mucins: a quantitative study of normal and diseased (KCS) tissue. Albrecht Von Graefes Arch Klin Exp Ophthalmol 215:209–221, 1981.
11. Hossmann KA, Huerter T, Oschlies U: The effect of dexamethasone on serum protein

extravasation and edema development in experimental brain tumors of cat. Acta Neuropathol 60:223–231, 1983.

12. Murphy GF, Harrist TJ, Bhan AK, Mihm MC Jr: Distribution of cell surface antigens in histiocytosis X cells. Quantitative immunoelectron microscopy using monoclonal antibodies. Lab Invest 48:90–97, 1983.

13. Gravekamp C, Koerten HK, Verwoerd NP, De Bruijn WC, Daems W Th: Automated image analysis applied to electron micrographs. Cell Biol Int Rep 6:656, 1982.

14. Gallyas F, Wolff JR, Boettcher H, Zaborsky L: A reliable and sensitive method to localize terminal degeneration and lysosomes in the central nervous system. Stain Technol 55:299–306, 1980.

15. Wolff JR, Eins S, Holzgraefe M, Zaborsky L: The temporo-spatial course of degeneration after cutting cortico-cortical connections in adult rats. Cell Tissues Res 214:303–321, 1981.

16. Horne RW: Special specimen preparation methods for image processing in transmission electron microscopy: a review. J Microsc 113:241–256, 1978.

17. Hayat MA: Principles and Techniques of Scanning Electron Microscopy. New York: Van Nostrand-Reinhold, 1974–1978.

18. Thiébaut F, Rigaut JP, Reith A: Improvement in specificity of the silver staining technique for AgNor-associated acidic proteins in paraffin sections. Stain Technol 59:181–185, 1984.

19. Henderson R, Unwin PNT: Three-dimensional model of purple membrane obtained by electron microscopy. Nature (Lond) 257:28–32, 1975.

20. Hawkes PW: Computer processing of electron micrographs. In: Principles and Techniques of Electron Microscopy, edited by Hayat, MA, volume 8, pp. 262–306. New York: Van Nostrand, 1978.

21. Frank J: Image analysis in electron microscopy. J Microsc 117:25–38, 1979.

22. Cole M: Instrument errors in quantitative image analysis. Microscope 19:87–103, 1971.

23. Harms H, Boseck S, Aus HM, Lenz V: Untersuchungen der Abtastbedingungen bei Zellbildern mit einem Mikroskop-TV-System. Microsc Acta 85:69–82, 1981.

24. Gerchberg RW, Saxton WO: Wave phase from image and diffraction plane pictures. In: Image Processing and Computer Aided Designs in Electron Optics, edited by Hawkes PW, pp. 66–81. London: Academic Press, 1973.

25. Burge RE, Dainty TC, Scott RF: Optical and digital image processing in high resolution electron microscopy. Ultramicroscopy 2:169–178, 1977.

26. Hawkins JK: Image processing principles and techniques. Adv Inf Syst Sci 3:113–214, 1970.

27. Kawata S, Ichioka Y, Suzuki T: Man-machine interactive image processing of high voltage electron micrograph images. Optik 52:235–246, 1979.

28. Gonzalez RC, Wintz P: Digital Image Processing. Reading, Mass.: Addison-Wesley, 1977.

29. Turin GL: An introduction to matched filters. IRE Trans Inf Theory IT6:311–329, 1960.

30. Frank J: Computer processing on electron micrographs. In: Advanced Techniques in Biological Electron Microscopy, edited by Koehler JK, pp. 215–274. Berlin: Springer, 1973.

31. Vollath D: The application of a normalized correlation function in image analysis. J Microsc 122:35–48, 1981.

32. Frank J: Reconstruction of non-periodic objects using correlation methods. In: Electron Microscopy 1978, edited by Sturgess JM, volume 3, part 3, Image Analysis, Proc 9th Int Congr EM, pp. 87–93. 1987.

33. Serra J: Image Analysis and Mathematical Morphology. London: Academic Press, 1982.

34. Skolnick MM, Sternberg SR, Neel JV: Computer programs for adapting two-dimensional gels to the study of mutation. Clin Chem 28:969–978, 1982.

35. Prewitt JMS: Object enhancement and extraction. In: Picture Processing and Psychopictorics, edited by Lipkin BS, Rosenfeld A, pp. 75–149. New York: Academic Press, 1970.
36. Rosenfeld A: Picture processing by computer. Comput Surv 1:147–176, 1969.
37. Duda RO, Hart PE: Pattern Classification and Scene Analysis. New York: Wiley, 1973.
38. Roberts LG: Machine perception of three dimensional solids. In: Optical and Electro-Optical Information Processing, edited by Tippett J, Berkovitz D, Clapp L, Koester C, Vanderburgh A, pp. 159–197. Cambridge, Mass.: MIT Press, 1965.
39. Favre A, Keller HJ, Mathieu O, Weibel ER: The use of local operations for pattern recognition on electron micrographs. Proc 5th Int Congr Stereol. Mikroskopie 37(Suppl):437–443, 1980.
40. Kemmer C, Voss K: Image processing in pathology: IV. Internal structure of mitochondria. Exp Pathol (Jena) 15:215–221, 1978.
41. Fram JR, Deutsch ES: On the quantitative evaluation of edge detection schemes and their comparison with human performance. IEEE Trans Comput C-24:616–628, 1975.
42. Taylor J, Bahr GF, Bartels PH, Bibbo M, Richards DL, Wied GL: Development and evaluation of automatic nucleus finding routines: thresholding of cervical cytology images. Acta Cytol 19:289–298, 1975.
43. Tanaka N, Ikeda H, Ueno T, Watanabe S, Imasato Y: Fundamental study of automatic cyto-screening for uterine cancer: II. Segmentation of cells and computer simulation. Acta Cytol 21:78–84, 1977.
44. Borst H, Abmayr W, Gais P: A thresholding method for automatic cell image segmentation. J Histochem Cytochem 27:180–187, 1979.
45. Cahn RL, Poulsen RS, Toussaint G: Segmentation of cervical cell images. J Histochem Cytochem 25:681–688, 1977.
46. Ullmann JR: Binarization using associative addressing. Pattern Recogn 6:127–135, 1974.
47. Panda DP, Rosenfeld A: Image segmentation by pixel classification in (gray level, edge value) space. IEEE Trans Comput C-27:875–879, 1978.
48. Robinson GS: Edge detection by compass gradient mask. Comput Graph Image Process 6:492–501, 1977.
49. Rosenfeld A, Thurston M: Edge and curve detection for visual scene analysis. IEEE Trans Comput C-20:562–569, 1971.
50. Abmayr W, Abele L, Kugler J, Borst H: Capabilities of a nonlinear gradient and a thresholding algorithm for the segmentation of Papanicolaou-stained cervical cells. Anal Quant Cytol 2:221–233, 1980.
51. Baky AA, Winkler DG, Hunter NR, Greenberg SD, Hodapp CJ, Kimzey SL: Nuclear boundary detection algorithm based on a minimax derivative statistic for atypical bronchial squamous epithelial cells. Anal Quant Cytol 3:33–38, 1981.
52. Beucher S, in Nawrath R, Serra J: Quantitative image analysis: theory and instrumentation. Microsc Acta 82:101–111, 1979.
53. Kovalewsky VA: Sequential optimization in pattern recognition and pattern description. Proc IFIP Congr, pp. 146–151. Amsterdam: North-Holland, 1968.
54. Montanari U: On the optimal detection of curves in noisy pictures. Commun ACM 14:335–345, 1971.
55. Martelli A: Edge detection using heuristic search methods. Comput Graph Image Process 1:169–182, 1972.
56. Kelly M: Edge detection by computer using planning. In: Machine Intelligence VI, pp. 397–409. Edinburgh: Edinburgh University Press, 1971.
57. Griffith AK: Mathematical models for automatic line detection. J Assoc Comput Mach 20:62–80, 1973.
58. Hueckel MJ: A local visual operator which recognizes edges and lines. J Assoc Comput Mach 20:634–647, 1973.

59. Davis LS: A survey of edge detection techniques. Comput Graph Image Process 5:248–270, 1975.
60. Weszka JS: A survey of threshold selection techniques. Comput Graph Image Process 7:259–265, 1978.
61. Meyer F: Iterative image transformations for an automatic screening of cervical smears. J Histochem Cytochem 27:128–135, 1979.
62. Lantuéjoul C: La squelettisation et son application aux mesures topologiques des mosaïques polycristallines. Paris: Thèse Ingénieur Ecole des Mines, 1978.
63. Beucher S, Lantuéjoul C: Use of watersheds in contour detection. International Workshop on Image Processing: Real Time Edge and Motion Detection Estimation, Doc CCETT, CTN/T/l/80, Doc IRISA numero 132, pp. 2/1–2/12. Rennes: CCETT, 1979.
64. Wynford-Thomas D, Garrahan N, Jasani B, Williams ED: Automated cell counting in tissue sections: a new approach by "multiple grey-level analysis." J Microsc 127:175–184, 1982.
65. Lester JM, Williams HA, Weintraub BA, Brenner JF: Two graph searching techniques for boundary finding in white blood cell images. Comput Biol Med 8:293–308, 1978.
66. Nawrath R, Serra J: Quantitative image analysis: applications using sequential transformation. Microsc Acta 82:113–128, 1979.
67. Ullman J: Pattern is in the eye of the beholder. New Scientist March 3:503–505, 1977.
68. Darling E Jr, Joseph R: Pattern recognition from satellite altitudes. IEEE Trans Syst Sci Cybern SCC4:38–47, 1968.
69. Fu KS: Digital Pattern Recognition. New York: Springer-Verlag, 1976.
70. Andrews HC: Computer Techniques in Image Processing. New York: Academic Press, 1970.
71. Krivenhov BE, Tverdokhleb PE, Chugui YV: Analysis of images by Hadamard optical transform. Appl Opt 14:1829–1834, 1975.
72. Andrews HC: Introduction to Mathematical Techniques in Pattern Recognition. New York: Wiley, 1972.
73. Han KS, McLaren RW, Lodwick GS: The application of image compression feature trans-generation techniques to the computer aided diagnosis of brain tumours. IEEE Trans Syst Man Cybern SMC-3:410–415, 1973.
74. Kanal LN, Randall NC: Recognition system design by statistical analysis. Proc 19th Natl Conf ACM, pp. D2.5-1–D2.5-10. New York: Association for Computing Machinery, 1965.
75. Uhr L, Vossler C: A pattern recognition program that generates, evaluates and adjusts its own operators. In: Pattern Recognition, pp. 349–364. New York: Wiley, 1966.
76. Batchelor BC: Pattern Recognition Ideas in Practice. New York: Plenum, 1976.
77. Cruttwell IA: Pattern recognition by automatic image analysis. Microscope 22:27–37, 1974.
78. Fukunaga K: Introduction to Statistical Pattern Recognition. New York: Academic Press, 1972.
79. Gelsema ES, Kanal LN: Pattern Recognition in Practice. Amsterdam: North-Holland, 1980.
80. Rosenfeld A, Kak AC: Digital Picture Processing. New York: Academic Press, 1976.
81. Hawkins JK: Textural properties for pattern recognition. In: Picture Processing and Psychopictorics, edited by Lipkin BS, Rosenfeld A, pp. 347–370. New York: Academic Press, 1970.
82. Yasnoff WA, Mui JK, Bacus JW: Error measures for scene segmentation. Pattern Recogn 9:217–231, 1977.
83. Gundersen HJG: Notes on the estimation of the numerical density of arbitrary profiles: the edge effect. J Microsc 11:219–223, 1977.

84. Miles RE: On the elimination of edge effects in planar sampling. In: Stochastic Geometry, edited by Harding EF, Kendall DG, pp. 228–247. New York: Wiley, 1974.
85. Lantuéjoul C: On the estimation of mean values in individual-analysis of particles. Microsc Acta Suppl 4:266–273, 1980.
86. Hougardy HP, Stienen HD: Edge correction in digital image analysis. Pract Metallogr Spec Issue 8:25–34, 1978.
87. Rigaut JP, Berggren P, Robertson B: Automated techniques for the study of lung alveolar stereological parameters with the IBAS image analyser on optical microscopy sections. J Microsc 130:53–61, 1983.
88. Paul J, Exner HE: Effectivity of detector and perimeter algorithms for automatic image analysis. Proc 3d Eur Symp Stereol, Stereol Iugosl 3(Suppl 1):189–198, 1981.
89. Rigaut JP, Margules S, Boysen M, Chalumeau MT, Reith A: Karyometry of pseudostratified, metaplastic and dysplastic nasal epithelia by morphometry and stereology. 1. A general model for automated image analysis of epithelia. Pathol Res Pract 174:342–356, 1982.
90. Schmeisser H: Stereological analysis in scanning electron microscopy. Microsc Acta 82:129–136, 1979.
91. Thiébaut F, Rigaut JP, Feren K, Reith A: Smooth-surfaced control and transformed C3H/10T-½ cells differ in cytology. A study by secondary electron, backscattered electron and image analysis. SEM 3:1249–1255. Chicago: SEM Inc. AMF O'Hare, 1984.
92. Boyde A, Reid SA: A new method of scanning electron microscopy for imaging biological tissues. Nature (Lond) 302:522–523, 1983.
93. Bauer B, Schwarz H, Nguyen Van Thanh: Digital image processing of combined SEM and EDX signals. J Microsc 130:325–330, 1983.
94. Jones AV, Smith KCA: Image processing for scanning microscopists. SEM 1:13–26. Chicago: SEM Inc. AMF O'Hare, 1978.
95. Jones AV: Computers and scanning microscopes: present and future trends. J Microsc Spectrosc Electron 5:591–601, 1980.
96. Lee RJ, Kelly JF: Overview of SEM-based automated image analysis. SEM 1:303–310. Chicago: SEM Inc. AMF O'Hare, 1980.
97. Pratt WK: Digital Image Processing. New York: Wiley, 1978.
98. Rosenfeld A: Digital Picture Analysis. Berlin: Springer-Verlag, 1976.
99. Fuchs H, Kedem ZM, Uselton SP: Optimal surface reconstruction from planar contours. Commun ACM 20:693–702, 1977.
100. Shantz MJ, MacCann GD: Computational morphology: three-dimensional computer graphics for electron microscopy. IEEE Trans Biomed Eng 25:99–103, 1978
101. Tenny JR, Long JW Jr, MacFarland WD, Vorbeck ML, Townsend JF, Martin AP: Computerized 3-dimensional analysis of mitochondrial structure. Comput Programs Biomed 12:1–6, 1980.
102. Moens PB, Moens T: Computer measurements and graphics of three-dimensional cellular ultrastructure. J Ultrastruct Res 75:131–141, 1981.
103. Perkins WJ, Green PJ: Three-dimensional reconstruction of biological sections. J Biomed Eng 4:37–43, 1982.
104. Gordon R, Herman GT: Three-dimensional reconstruction from projections: a review of algorithms. Int Rev Cytol 38:111–151, 1974.
105. Bauer B, Exner HE: Quantification and reconstruction of spatial objects and surfaces by computer-aided stereometry. Proc 3d Eur Symp Stereol, Stereol Iugosl 3(Suppl 1):255–262, 1981.
106. Williams RC, Wyckoff RWG: The thickness of electron-microscopic objects. J Appl Phys 15:712–717, 1944.
107. Horn BKP: Shape from shading: a method for obtaining the shape of a smooth opaque object from one view. Cambridge: PhD Thesis, MIT, 1970.

108. Howell PGT: Stereometry as an aid to stereological analysis. J Microsc 118:217–220, 1980.
109. Kristensen SEL, Papadimitriou JM: The measurement of cell volume and surface area by SEM photogrammetry. J Microsc 124:155–161, 1981.
110. Boyde A: Photogrammetry of stereo pair SEM images using separate measurements from the two images. SEM 1:101–115. Chicago:IITRI, 1974.
111. Howell PGT, Boyde A: The use of an x-y digitiser in SEM photogrammetry. Scanning 3:218–219, 1980.
112. Lane GS: The application of stereographic techniques to the scanning electron microscope. J Sci Instrum Ser 2 2:565–572, 1969.
113. Hilliard JE: Quantitative analysis of scanning electron micrographs. J Microsc 95:45–58, 1972.
114. Underwood EE: Quantitative shape parameters for microstructural features. Microscope 24:49–64, 1976.
115. Medalia AI: Dynamic shape factors of particles. Powder Technol 4:117–138, 1970–1971.
116. Young IT, Walker JE, Bowie JE: An analysis technique for biological shape. Inf Control 25:357–370, 1974.
117. Bookstein FL: The Measurement of Biological Shape and Shape Change. Lecture Notes in Biomathematics 24. Berlin: Springer-Verlag, 1978.
118. Cruz-Orive LN: Quantifying ''pattern'': a stereological approach. J Microsc 107:1–18, 1976.
119. Lantuéjoul C, Maisonneuve F: Geodesic methods in quantitative image analysis. Pattern Recogn 17:177–187, 1984.
120. Ferrario VF, Vizzotto L, Molinari Tosatti MP, Miani A: The evaluation of anisotropy in oriented structures: a statistico-mathematical approach. J Submicrosc Cytol 14:491–498, 1982.
121. Mandelbrot BB: The Fractal Geometry of Nature. San Francisco: Freeman, 1982.
122. Rigaut JP: An empirical formulation relating boundary lengths to resolution in specimens showing non-ideally fractal dimensions. J Microsc 133:41–54, 1984.
123. Paumgartner D, Losa G, Weibel ER: Resolution effects on the stereological estimation of surface and volume and its interpretation in terms of fractal dimensions. Stereology 5, Proc 5th Int Congr Stereol. J Microsc 121:51–63, 1981.
124. Rigaut JP, Berggren P, Robertson B: Resolution-dependence of stereological estimations: interpretation, with a new fractal concept, of automated image analyser-obtained results on lung sections. Proc 6th Int Congr Stereol. Acta Stereol 2(Suppl 1):121–124, 1983.
125. Julesz B: Visual pattern discrimination. IRE Trans Inf Theory IT-8:84–92, 1962.
126. Haralick RM, Shanmugam K, Dinstein I: Textural features for image classification. IEEE Trans Syst Man Cybern SMC3:610–621, 1973.
127. Pressman NJ: Markovian analysis of cervical cell images. J Histochem Cytochem 24:138–144, 1976.
128. Rigaut JP: Image analysis in electron microscopy. In Electron Microscopy in Human Medicine, edited by Johannessen JV, volume 11b, pp. 197–231. Maidenhead: McGraw-Hill International, 1983.
129. McNutt NS, Cain WR: Quantitative electron microscopic comparison of lymphocyte nuclear contours in mycosis fungoides and in benign infiltrates in skin. Cancer 47:698–709, 1981.
130. Payne CM, Nagle RB, Lynch PJ: Quantitative electron microscopy in the diagnosis of mycosis fungoides. A simple analysis of lymphocytic nuclear convolutions. Arch Dermatol 120:63–75, 1984.
131. Johnson PW, Barnett RI, Mackninon EA: Image analysis of the acetic-alcohol fixed metaphase chromosome. J Submicrosc Cytol 14:31–43, 1982.
132. Chegini N, Hilder VA, Gregory SP, Maclean N: Structural transitions of chromatin in

isolated *Xenopus* erythrocyte nuclei. II. Computer based image analysis. J Submicrosc Cytol 13:309–319, 1981.

133. Nicolini C: Chromatin structure: from nuclei to genes (review). Anticancer Res 3:63–86, 1983.

134. Nagl W: Species and hybrid diagnosis in plants by means of quantitative light and electron microscopic morphometry of chromatin texture. Microsc Acta Suppl 4:19–25, 1980.

135. Burger G: Anwendung bildanalytischer Verfahren in der Zytologie, Ergebnisse und Program. Morphometrie und Stereologie in der Pathologie Workshop, Neuherberg, 1984.

136. Altmann GG, Leblond CP: Changes in the size and structure of the nucleolus of columnar cells during their migration from crypt base to villus top in rat jejunum. J Cell Sci 56:83–99, 1982.

137. Kunze KD, Dimmer V, Haroske G: Differentiation of human T and B periphereal blood lymphocytes by high resolution cell image analysis. Anal Quant Cytol 5:167–172, 1983.

138. Voss K, Kemmer C: Image processing in pathology. VIII. Internal structure of mitochondria during treatment of HLP-patients with Regadrin. Exp Pathol 15:311–318, 1978.

139. Ohno S: Stereological application of thick sections to determination of size distribution of spherical cell organelles by high-voltage electron microscopy. J Electron Microsc 29:98–105, 1980.

140. Bradbury S: Microscopical image analysis: problems and approaches. J Microsc 115:137–150, 1979.

141. Ockleford CD, de Voy K, Hall HMK: Image analysis of electron micrographs of isolated negatively stained human placental coated vesicles. Cell Int Rep 4:810, 1980.

142. Petroutsos G, Patey A: Sclérocornée. Etude ultrastructurale et morphologique. J Fr Ophthalmol 6:769–775, 1983.

143. Beasley WL, Holland GR: A quantitative analysis of the innervation of the pulp of the cat's canine tooth. J Comp Neurol 178:487–494, 1978.

144. Diemer NH: Quantitative morphological studies of neuropathological changes. Part 1. CRC Crit Rev Toxicol 10:215–263, 1982.

145. Eins S, Wilhelms E: Assessment of preparative volume changes in central nervous tissue using automatic image analysis. Microscope 24:29–38, 1976.

146. Holland GR: The effect of buffer molarity on the size, shape, and sheath thickness of peripheral myelinated nerve fibres. J Anat 135:183–190, 1982.

147. Kensler RW, Stewart M: Frog skeletal muscle thick filaments are three-stranded. J Cell Biol 96:1797–1802, 1983.

148. Squire J, Edman AC, Freundlich A, Harford J, Sjoeström M: Muscle structure, cryo-methods and image analysis. J Microsc 125:215–225, 1982.

149. Laoussadi S, Dupoisot H, Constans A, Daury G, Delbarre F: Morphologie cristalline des dépots de trihexosylcéramides caractéristiques de la maladie de Fabry (forme rhumatismale notamment). CR Soc Biol (Paris) 175:55–67, 1981.

150. Barton SP, Marks R: Measurement of collagen-fibre diameter in human skin. J Cutaneous Pathol 11:18–26, 1984.

151. Larsson HO, Lorentzon R, Boquist L: Structure of the parathyroid glands, as revealed by different methods of fixation. A quantitative light and electron microscopic study in untreated Mongolian gerbils. Cell Tissue Res 235:51–58, 1984.

152. Weibel ER: Stereological methods in cell biology: where are we—where are we going? J Histochem Cytochem 29:1043–1052, 1981.

153. Kraehenbuhl JP, Racine L, Griffiths GW: Attempts to quantitate immunocytochemistry at the electron microscope level. Histochem J 12:317–332, 1980.

154. Kraemer M, Vassy J, Foucrier J, Rigaut JP, Chalumeau MT: Subcellular localization of transferrin and albumin synthesis in the same rat hepatocyte by immunoenzymatic, immunoferritin and image analysis methods. Biol Cell 40:103–108, 1981.

155. Walker DG: Cationic ferritin binding sites and surface charge densities of transformed cells. J Histochem Cytochem 29:255–265, 1981.
156. Capowski JJ, Cruce WLR: How to configure a computer-aided neuron reconstruction and graphics display system. Comput Biomed Res 12:569–587, 1979.
157. Macagno ER, Levinthal C, Sobel I: Three-dimensional computer reconstruction of neurons and neuronal assemblies. Annu Rev Biophys Bioeng 8:323–351, 1979.
158. Mazziota JC, Hamilton BL: Three-dimensional computer reconstruction and display of neuronal structure. Comput Biol Med 7:265–279, 1977.
159. Johnson DL: Automated scanning electron microscopic characterization of particulate inclusions in biological tissues. SEM 3:1211–1228. Chicago: SEM Inc. AMF O'Hara, 1983.
160. Tawashi R: Size-shape analysis of calcium oxalate crystals in the study of stone formation. SEM 1: 397–406. Chicago: SEM Inc. AMF O'Hare, 1983.
161. Stettler LE, Groth DH, Platek SF: Automated characterization of particles extracted from human lungs: three cases of tungsten carbide exposure. SEM 1:439–448. Chicago: SEM Inc. AMF O'Hare, 1983.
162. Boysen M, Reith A: Surface structure in normal, metaplastic and dysplastic nasal mucosa of nickel workers. A SEM and post-SEM histopathological study. SEM 3:35–42. Chicago: SEM Inc. AMF O'Hare, 1980.
163. Roels F, Pauwels M, Cornelis A, Kerckaert I, Van der Spek P, Goovaerts G, Versieck J, Goldfischer S: Peroxisomes (microbodies) in human liver; cytochemical and quantitative studies of 85 biopsies. J Histochem Cytochem 31:235–237, 1983.
164. Jaulin R: Jeux et Jouets. Paris: Aubier, 1972.
165. Debord G: La Société du Spectacle. Paris: Buchet-Chastel, 1967.
166. Weibel ER: Stereological Methods, volume 1, Practical Methods for Biological Morphometry. London: Academic Press, 1979.
167. Laing R: The Politics of Experience and the Bird of Paradise. London: Tavistock, 1967.
168. Kant E: Critique de la Raison Pure (1781), Textes Choisis. Paris: PUF, 1958.
169. McCrone WC, quoted in Underwood EE: Stereology in automatic image analysis. Microscope 22:69–80, 1974.

part 7

a worked example

This final chapter by R. Østerby presents a worked example, showing how to use the results of a pilot study to predict a more optimal (i.e., more cost-beneficial) sampling regime. It illustrates how the volume of glomerular basement membrane in rat kidney can be estimated efficiently by taking the reader through all intermediate calculations. Exactly similar calculation sequences could be used with any of the other component densities (surface, length, number) described in earlier chapters of this book.

chapter 14

a stereological study in practice

R. Østerby

INTRODUCTION

At the start of a stereological study it is often advantageous to perform a pilot experiment. Such an experiment forms the basis of the further design of the study in either of two ways:

1. By making a complete analysis of the results at all levels (nested analysis of variance) and deducing from this at which levels and to what extent the design should be changed to be optimal, i.e., more efficient at the lowest possible cost.[1-3]
2. By looking at the data, especially the variation at each level, and using common sense to see where the design should be changed. In doing this, a few simple general rules of thumb should be applied:
   (a) Pay most attention to the variation at the highest level of sampling (between animals) and least to that at the lowest level (e.g., point counting).
   (b) Obtain the points counted from a large number of tissue blocks sampled throughout the organ rather than from several micrographs from one or two blocks. The total number of points should never be concentrated within a few micrographs.
   (c) For the number of points (P) hitting the structure of interest (α), $P(\alpha)$, use a *total* of about 100 per animal. This suffices in most, if not all, cases.
   (d) Make the test system such that the numerator and the denominator in the ratio estimator (e.g., V_V) are of the same order of magnitude. For example, for a volume fraction of ~0.06 use a 1 : 16 grid, for V_V ~0.25 a 1 : 4 grid.

Although these considerations are illustrated with reference to volume densities estimated by point counting, they are also relevant to surface, length, and numerical densities estimated by intersection and profile counting methods.

DECISIONS FOR THE PILOT EXPERIMENT

The design can usually be based on some a priori knowledge of the structures in question. The unknown in the whole experiment is usually the variation among animals in the groups under study. This variation, together with the difference in mean values between groups, determines whether there is a statistically significant difference between the groups.

A practical example is given below concerning the estimation of a structural quantity, the total volume (V) of glomerular basement membrane (BM) material, V(BM), in groups of diabetic (D) and control (C) rats. The study requires sampling at three different microscopic levels: light microscopy and electron microscopy at low ($\sim 2000 \times$) and higher ($\sim 12,000 \times$) magnification.

The final estimate of total BM volume is

$$V(\text{BM}) = V_V(\text{BM/tuft}) \times V_V(\text{tuft/glom}) \times V_V(\text{glom/kidney}) \times V(\text{kidney})$$

where V(kidney) is the volume of the reference space, obtained by weighing or fluid displacement. If this is not possible, use the very efficient Cavalieri's principle.[4] For this purpose the organ is sectioned at equidistant intervals and the areas of cut surfaces are measured (by point counting), added together, and multiplied by the distance between sections. On the order of 10 cuts per organ are usually enough. No assumptions regarding the shape of the organ need be made, and the direction may be chosen solely for practical reasons. The distance between sections must be known. (For further details see Gundersen.[4]) V_V(glom/kidney) is estimated by light microscopy, level I; V_V(tuft/glom) is estimated by electron microscopy (EM), $2000 \times$ magnification, level II; and V_V(BM/tuft) is estimated by EM, $12,000 \times$ magnification, level III.

In our pilot experiment we studied three animals in each group. The minimum number obviously is two. One may also choose to start with five in each group. The final answer may be at hand at the end of the pilot experiment! (The five animals must, of course, be selected from the larger group by unbiased sampling, a requirement that must be fulfilled at each sampling level.)

On a systematic (equidistant, independently positioned) set of sections (four to six) through the perfusion-fixed kidney, an independent set of visual fields is projected to the table, and V_V(glom/kidney) is estimated with a 1 : 19 grid (see Fig. 1). The number of sections and number of visual fields that are necessary depend on the organ heterogeneity with respect to the phase of interest at that level.

For each animal three glomerular profiles (sampling, see Ref. 5) from three different blocks are photographed in toto. V_V(tuft/glom) is estimated with a 1 : 4 grid (see Fig. 2).

From each of the three glomerular profiles, a sample (sampling, see Ref. 6) of micrographs at $12,000\times$ is obtained, covering about one-fourth of the total area, and V_V(BM/tuft) is estimated applying a 1 : 4 grid (see Fig. 3).

RESULTS

Tables 1–3 present the results of the pilot study. These results are the basis for an analysis of the pilot study design, which is used to choose a more optimal design for future studies on the same kind of material.

EVALUATION OF RESULTS OF THE PILOT STUDY

The variation in V_V(BM/tuft) between animals (Table 1), between glomeruli (Table 2), and between micrographs (Table 3) is examined. Examples of calculating the variance at three levels are shown below. A more detailed and fully calculated variance analysis can be found in the references.[1-3]

Variation between Animals (*a*) (Table 1)

Mean $R = V_V$(BM/tuft) $= 0.127$ (controls)

Observed coefficient of error: $\mathrm{OCE}_a(R) = 0.0842$

Observed variance (including that from lower levels): $\mathrm{OS}_a^2(R) = 0.00034$
$[(\sqrt{3} \times 0.0842 \times 0.127)^2 = 0.00034]$

How much of this variation is due to variation at lower levels, e.g., between glomeruli?

Variation between Glomeruli (*g*) (Table 2)

Rat	$\mathrm{OCE}_g(R)$	$\mathrm{OS}_g^2(R)$
1	0.0681	0.00022
2	0.136	0.00118
3	0.0561	0.00011
		0.00151

$$\mathrm{OS}_g^2(R) = \frac{0.00151}{3} = 0.00050$$

Conclusion: the variation between glomeruli in this case is large compared to that between animals.

The variation between animals, based on only three rats, is of course estimated with considerable uncertainty. However, extra glomeruli from the em-

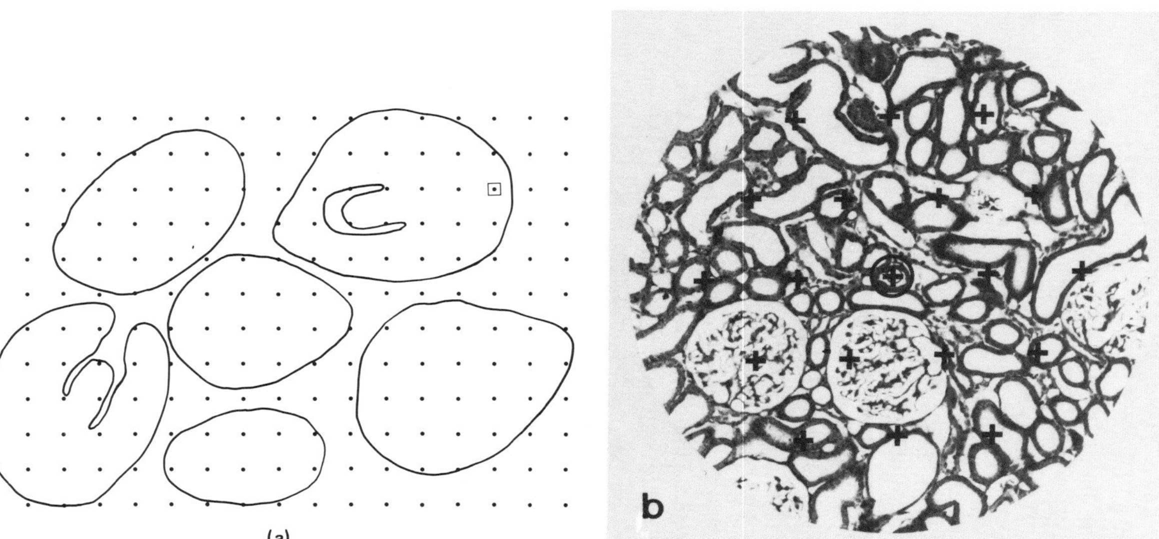

FIGURE 1. Light microscopy. Six equidistant transections were cut through the perfusion-fixed rat kidney, giving seven slabs of tissue. The six slabs with a "right-hand" cut surface were embedded in paraffin in one capsule, and one section was stained for light microscopy (a). At each point (sampling, see Ref. 5) the visual field is projected to the table, i.e., a *systematic random sample* throughout the kidney is obtained. V_V(glom/kidney) is estimated with a 1 : 19 grid (b).

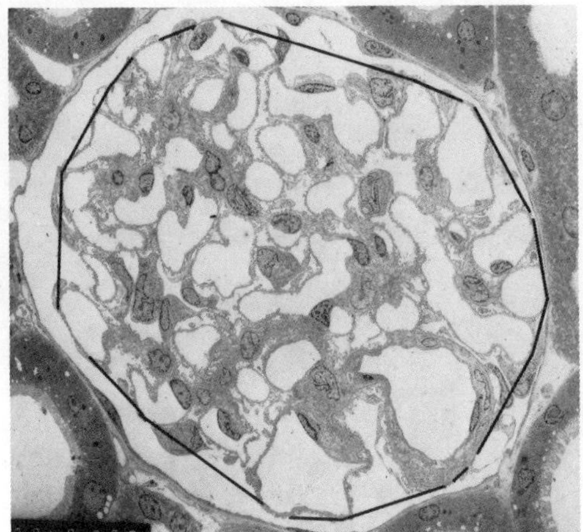

FIGURE 2. Low-magnification electron micrograph of a glomerular cross section. If necessary, more micrographs are mounted together in a montage. The circumscribed minimal polygon corresponds to the definition used at light microscopy in estimating V_V(glom/kidney). At this level V_V(tuft/glom) is estimated with a 1 : 4 grid (the tuft is the capillary network, delineated by epithelial cells). The fields are those sampled for higher-magnification EM (Fig. 3).

bedded blocks are easily available, and we decide to use five glomeruli per animal in the final study.

How much of the variation between glomeruli is due to the variation between micrographs?

Variation between Micrographs (m) (Table 3)

Rat	Glomerulus	No. of micrographs	$OCE_m(R)$	$OS^2_m(R)$		
	1	17	0.164	0.00846		
1	2	13	0.240	0.01325		
	3	10	0.249	0.00737		
				Σ 0.02908	mean = 0.00969	
	1	10	0.163	0.00358		
2	2	13	0.127	0.00642		
	3	12	0.138	0.00521		
				Σ 0.01521	mean = 0.00507	
	1	14	0.157	0.00433		
3	2	11	0.200	0.00406		
	3	12	0.186	0.00578		
		Mean 12		Σ 0.01417	mean = 0.00472	

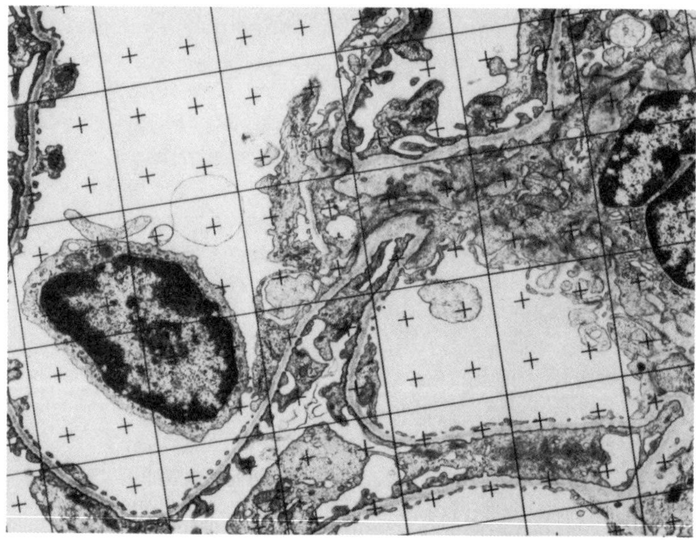

FIGURE 3. V_V(BM/tuft) estimated with a 1 : 4 grid. The tuft is identified exactly as at the preceding level. Conformity in definitions at sequential levels in the hierarchical study is a requirement for the final calculation of V(BM) to be valid.

$$OS_m^2(R) = \frac{0.00969 + 0.00507 + 0.00472}{3} = 0.00649$$

$$S_g^2(R) = OS_g^2(R) - OS_m^2(R) = 0.00050 - \frac{0.00649}{12}$$
$$= 0.00050 - 0.000541$$

Conclusion: The variation between micrographs (including the measurement itself, point counting) is of the same order as that between glomeruli.

If at this level we find the variation between micrographs too large, it can be due either to a true large variation between micrographs (i.e., take more micrographs) or to the point counting. This can, of course, be tested [e.g., to determine the effect on $CE_m(R)$ of increasing the number of points, double the number by two independent placings of the test screen on each micrograph]. However, the rule of thumb tells us that a total of 100 points per animal [$P(\alpha)$] usually, perhaps always, is enough.

About 10 micrographs per glomerulus suffice. In this case we will stick to the sampling protocol in photographing at 12,000× magnification.

TABLE 1. Estimation of Total Volume (V) of Glomerular Basement Membrane (BM) in Kidneys of Individual Control and Diabetic Rats: Overall Results, Including All Levels[a]

Rat	V(kidney) (mm³)	V_v(glom/kidney) (19 : 1)			V_v(tuft/glom) (4 : 1)			V_v(BM/tuft) (4 : 1)			V(BM) in one kidney (mm³)
		$P(\alpha)$	$P(r)^b$	$\hat{R}$	$P(\alpha)$	$P(r)^b$	$\hat{R}$	$P(\alpha)$	$P(r)^b$	$\hat{R}$	
Controls											
1	885	343	388	0.047	398	167	0.60	143	2.83	0.126	3.145
2	852	437	390	0.069	375	137	0.68	159	273	0.146	4.991
3	805	370	379	0.051	279	102	0.68	104	238	0.109	3.043
Mean	847			0.052			0.65			0.127	3.726
CE	0.027			0.067			0.041			0.084	0.170
Diabetic											
4	1052	487	519	0.049	399	158	0.63	148	362	0.102	3.312
5	1006	530	467	0.060	326	139	0.59	170	261	0.163	5.805
6	1045	588	560	0.055	248	98	0.63	188	361	0.130	4.707
Mean	1034			0.055			0.62			0.132	4.608
CE	0.014			0.058			0.022			0.134	0.157

[a] $P(\alpha)$, Points falling on object of interest; $P(r)$, points falling on reference space; $\hat{R}$, ratio $P(\alpha)/4P(r)$ or $P(\alpha)/19P(r)$.

[b] Multiply by the test grid factor to get the same conditions as for $P(\alpha)$; for a 1 : 4 grid, for example, the observed $P(r)$ must be multiplied by 4 to maintain the same numerical weighting of points as for $P(\alpha)$.

TABLE 2. Individual Glomeruli: Estimation of Fractional Volume of Glomerular Basement Membrane, V_VBM/glom

Rat	Glomerulus	4 : 1			$r(P(\alpha), P(r))$[b]	CE(R)[c]
		$P(\alpha)$	$P(r)$	$\hat{R}$[a]		
1	1	48	88	0.136		
	2	55	103	0.133		
	3	40	92	0.109		
	Point totals	143	283	0.126	0.680	0.0681
2	1	51	110	0.116		
	2	70	100	0.175		
	3	38	63	0.151		
	Point totals	159	273	0.146	0.671	0.136
3	1	38	85	0.112		
	2	26	68	0.096		
	3	40	85	0.118		
	Point totals	104	238	0.109	0.991	0.0561
4	1	49	92	0.133		
	2	53	141	0.094		
	3	46	129	0.089		
	Point totals	148	362	0.102	0.314	0.116
5	1	79	95	0.208		
	2	26	77	0.084		
	3	65	89	0.183		
	Point totals	170	261	0.163	0.997	0.219
6	1	65	148	0.110		
	2	47	75	0.157		
	3	76	138	0.138		
	Point totals	188	361	0.130	0.872	0.098

[a] $\hat{R} = P(\alpha)/4P(r)$.
[b] r = correlation coefficient.
[c] Coefficient of error of three glomeruli.

Point Counting

On average we counted 150 points for each of the six animals for BM $[P(\alpha)]$ (see Table 2). Since the contribution of the variation between micrographs to the total is low, we should aim at having at most about 100 points per animal. This means we have to increase the test point area by a factor of $150/100 = 1.50$ (or the distance between the test points by a factor of $\sqrt{1.50} = 1.23$). In addition, we must consider that the total area under study will be increased by a factor $5/3$ (five glomeruli instead of three; see above). In the new grid, therefore, the distance between fine points is increased by a factor of $\sqrt{(150/100) \times (5/3)} = 1.59$, or the number of previous fine points per micrograph is reduced by the factor $\sim(100/150) \times (3/5) = 0.40$; that is, make the new grid with less than half the original number of points.

The ratio R between points falling on the basal membrane and the reference space, the tufts, $P(\alpha)/P(r)$, is on the order of $0.10-0.15$ (see Table 3), and we therefore change to a 1 : 9 grid (as illustrated in Fig. 4), instead of the 1 : 4 grid in Fig. 3. When placing this new grid on the micrographs we must again take into consideration the point reduction of 0.4, i.e., increase the point distance by 1.6 (see above) by choosing the correct magnification of the grid.

By using the above methods, we will achieve a more efficient test system. However, experience has shown that most often the variation found at the micrograph level does not contribute very much to the total variation. To obtain maximum efficiency, it is most important after the pilot study to redesign the subsequent stereological studies with respect to the optimal number of animals and number of blocks.

FORMULAS AND TERMINOLOGY

Formulas

Coefficient of variation: $\mathrm{CV}(x) = \mathrm{SD}(x)/\bar{x}$

Variance: $s^2(x) = \mathrm{SD}^2(x) = [\mathrm{CV}(x)\bar{x}]^2$

Coefficient of error: $\mathrm{CE}(x) = \mathrm{CV}(x)/\sqrt{n}$ [i.e., from $\mathrm{CE}(x)$ to $s^2(x) = (\sqrt{n}\,\mathrm{CE}(x)\bar{x})^2$]

Ratio estimators: $R = y/x$, which in this example $= P(\alpha)/P(r)$

$\mathrm{CE}(R) = [\mathrm{CE}^2(x) + \mathrm{CE}^2(y) - 2r(x, y)\mathrm{CE}(x)\mathrm{CE}(y)]^{1/2}$

Ratio of means $= \bar{y}/\bar{x} = \Sigma^n y/\Sigma^n x$ estimated from, say, the point totals on n micrographs

Correlation coefficient $= r$

Mean $= \bar{x}$

Terminology

$\mathrm{OCE}(R) = $ observed coefficient of error

$\mathrm{OS}^2(R) = $ observed variance of ratio R

TABLE 3. Estimation of Volume Fraction of Basement Membrane per Tuft, V_V(BM/tuft): Individual Micrographs, Point Counting Example, Rat No. 1 (Level III)

Glomerulus	Micrograph	4:1		$\hat{R}^a$	$r(P(\alpha), P(r))$	CE(R)
		$P(\alpha)$	$P(r)$			
1	1	6	6			
	2	3	1			
	3	4	5			
	4	4	13			
	5		1			
	6		1			
	7	2	9			
	8	4	11			
	9	3	5			
	10	5	5			
	11	4	7			
	12	1	2			
	13	1	2			
	14	6	6			
	15	1	4			
	16	2	7			
	17	3	2			
	Point totals	48	88	0.136	0.503	0.164
2	1	2	10			
	2	8	7			
	3	4	10			
	4	2	8			
	5	5	6			
	6		8			
	7	9	10			
	8		15			
	9	4	6			
	10	10	12			
	11	3	5			
	12	4	4			
	13	4	2			
	Point totals	44	103	0.133	−0.006	0.240
3	1	1	12			
	2	4	13			
	3	8	9			
	4	3	11			
	5		7			
	6	8	5			
	7	6	12			
	8	3	9			
	9	3	9			
	10	4	5			
	Point totals	40	92	0.109	−0.189	0.249

$^a \hat{R} = P(\alpha)/4P(r)$.

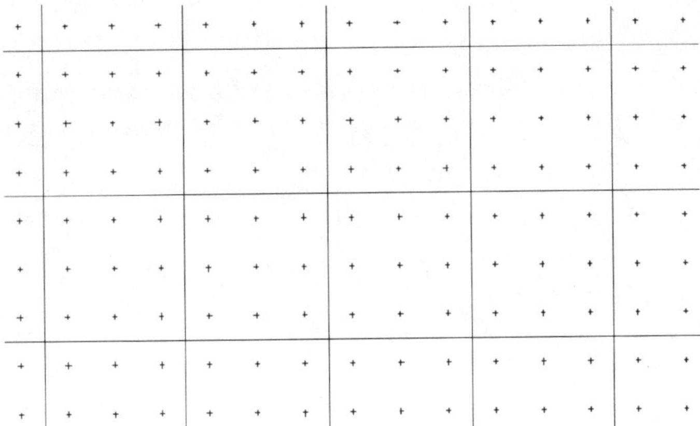

FIGURE 4. Test screen with fine and coarse points in 1 : 9 ratio.

$S^2(R)$ = "true variance" = $OS^2(R)$ minus variance from lower levels
Subscripts: a, animals; g, glomeruli; m, micrographs

REFERENCES

1. Gundersen HJG, Østerby R: Optimizing sampling efficiency of stereological studies in biology: or "Do more less well!" J Microsc 121:65, 1980.
2. Nicholson WL: Application of statistical methods in quantitative microscopy. J Microsc 113:223, 1978.
3. Sokal RR, Rohlf FJ: Nested analysis of variance. In: Biometry: The Principles and Practice of Statistics in Biological Research, Chapter 10. San Francisco: Freeman, 1981.
4. Gundersen HJG: Stereology of arbitrary particles. J Microsc 143:3, 1986.
5. Kroustrup JP, Gundersen HJG: Sampling problems in an heterogeneous organ. J Microsc 132:43, 1982.
6. Østerby R, Gundersen HJG: Sampling problems in the kidney. In: Lecture Notes in Biomathematics, edited by Miles RE, Serra J, p. 185. Berlin: Springer, 1978.

nomenclature

a	area of *individual* profile
A	area of profiles
A_A	areal density
A_T	test area
b	boundary length or perimeter of *individual* profile
B	boundary length of profiles
B_A	boundary length density on unit area
d	diameter of profile
d	spacing of test points
D	diameter (of spherical particles)
h	caliper diameter of plane figure
H	caliper diameter (tangent diameter) of solid
$\overline{H}$	mean caliper diameter
I	intersections between test line and surface or boundary
I_L	intersection density on test line
k	constant or coefficient
l	length of intercept or chord
$\overline{l}$	mean linear intercept
L	length of line or intercept (straight line)
L_L	length density of intercepts on test line
L_T	length of test line
n	number of profiles
n	number of samples in statistics
N	number of discrete objects or profiles
N_A	numerical density of profiles on unit test area
N_L	numerical density of intercepts on test line

N_N	relative number
N_V	numerical density of particles in unit volume
P	test point
P_T	test point set
P_P	test point density
Q	transection of spatial lineal features with plane
Q_A	transection density on unit area
q	lattice ratio for coherent test systems
r	radius of circle
R	radius of sphere
S	surface area
S_V	surface density in unit volume
S_S	relative surface
SD	standard deviation
SE	standard error of the mean
t	section or slice thickness
T	subscript designating test system
v	volume of *individual* object
V	volume of components or structure
V_V	volume density
β	shape-dependent coefficient
σ^2	variance

The following symbols follow a proposal by Weibel et al. whereby (a) capital letters refer to collective treatment of structure and lowercase letters to individual structures (e.g., v = volume of individual; V = total volume of group of structures) and (b) in subscripts, capital letters identify reference of test system and lowercase letters identify first the object and second the reference system. For example,

Individual dimensions:

   v_i (e.g., v_{mi}, volume of a mitochondrion)

   a_i (e.g., a_{mi}, transection or profile area of a mitochondrion)

Collective dimensions of a group of structures:

   V_i (e.g., V_{li}, volume of the liver; V_{mi}, volume of the chondriome)

Relative dimensions:

   V_{Vi} volume fraction of object i in volume of reference space

   (e.g., $V_{Vmi,par}$, fraction of mitochondria per volume parenchyma; $V_{Vmi,cyt}$, fraction of mitochondria per volume cytoplasm)

glossary

Anisotropy Exhibiting different properties in different directions of space.

Arithmetic mean Sum of sample values divided by number in sample (= average).

Average Sum of sample values divided by number in sample (= arithmetic mean).

Biased estimate Estimate that deviates from the true value by a certain amount and in a certain direction.

Boundary Outline of a profile (also called perimeter or circumference).

Coefficient of variation (CV) Ratio of standard deviation (SD) to mean (m). CV (%) = SD/m × 100.

Containing space Structure space to which parameters describing internal components are related.

Density Quantity per unit volume, unit area, unit surface, unit length, or unit number.

Harmonic mean The reciprocal of the sum of reciprocals of sample values divided by sample number.

Intercept Segment or chord of test line contained within object and extending between two points on the object's surface.

Intersection Point where test line crosses an object surface or its membrane trace.

Isotropy Having the same properties in all directions of space.

Morphometry Quantitative morphology; the measurement of structures by any method, including stereology.

Object Discrete element of structure, e.g., organelle.

Orientation Direction in space.

Particle Discrete element of structure, e.g., nucleus or mitochondrion.

Perimeter Outline of a profile (or boundary, or circumference).

Probe Geometrc entity (plane, line, point, slice, etc.) used to measure a structure.

Profile Sharply outlined plane figure resulting from sectioning (or projecting) an object with a plane (onto a plane).

Random sample Sample obtained by random process (e.g., by means of random numbers).

Random section Section of structure obtained by random sampling, whereby all regions of the structure and all orientations have the same chance of being cut.

Reference system Quantity to which component is related (containing space, test system, etc.).

Section Plane intersecting a structure; sometimes thin slice of thickness t.

Slice ''Section'' of positive thickness t.

Stereology Set of mathematical methods relating three-dimensional parameters of structure to measurements obtainable on sections.

Test system Set of geometric elements (points, lines, areas) used to obtain stereological estimates.

bibliography

JOURNALS

Acta Stereologica (Ljubljana). Published on behalf of the International Society for Stereology.

Analytical Cellular Pathology (Elsevier, Amsterdam). Published on behalf of the European Society for Analytical Cellular Pathology.

Analytical and Quantitative Cytology and Histology (Chicago).

Journal of Microscopy (Blackwell's, Oxford). Official journal of the International Society for Stereology, published with the Royal Microscopical Society.

Laboratory Investigation (Baltimore).

Microscopica Acta (Stuttgart).

Mikroskopie (Wien).

PROCEEDINGS

Haug H (ed.): Proceedings 1st International Congress for Stereology Vienna. Wien: Congressprint, 1963.

Elias H (ed.): Stereology. Proceedings 2d International Congress for Stereology. New York: Springer-Verlag, 1967.

Weibel ER, Meek G, Ralph B, Echlin P, Ross R (eds.): Stereology 3. Proceedings 3d International Congress for Stereology. Oxford: Blackwell Scientific, 1972.

Underwood EE, de Wit R, Moore GA (eds.): Proceedings 4th International Congress for Stereology. National Bureau of Standards Special Publication No. 431. Washington, D.C.: U.S. Government Printing Office, 1976.

Adam H, Bernroider G, Haug H (eds.): Proceedings 5th International Congress for Stereology. Mikroskopie (Wien), supplement 37. Wien: Verlag Georg Fromme, 1979.

Kalisnik M (ed.): Contemporary Stereology. Proceedings 3d European Symposium for Stereology. Stereologia Iugoslavia, volume 3, supplement 1. Ljubljana, 1981.

Kalisnik M (ed.): Acta Stereologica. Proceedings 6th International Congress for Stereology. Acta Stereologica, volume 2, supplement 1. Ljubljana, 1983.

Kalisnik M, Karlsson B, Warren R, Wasén J (eds.): Proceedings 4th European Symposium for Stereology. Acta Stereologica, volume 5. Ljubljana, 1986.

BOOKS

DeHoff RT, Rhines FN: Quantitative Microscopy. New York: McGraw-Hill, 1968.

Williams MA: Quantitative Methods in Biology. Oxford: North-Holland, 1977.

Weibel ER: Stereological Methods, volume 1, Practical Methods for Biological Morphometry. London: Academic Press, 1979.

Aherne WA, Dunnill MS: Morphometry. London: Arnold, 1982.

Elias H, Hyde DM: A Guide to Practical Stereology. Basel: Karger, 1983.

SOCIETIES

European Society for Analytical Cellular Pathology, % Dr. G. Burger, GSF München, Ingolstädter Landstr. 1, D-8042 Neuherberg, Federal Republic of Germany.

International Society for Stereology, % Dr. T. Mattfeldt, Secretary/Treasurer, Institute of Pathology, Im Neuenheimer Felt 220/221, D-6900 Heidelberg, Federal Republic of Germany.

index